Fundamentals of Agronomy

NIPA® GENX ELECTRONIC RESOURCES & SOLUTIONS P. LTD.

New Delhi-110 034

Fundamentals of Agronomy

B.S. Lalitha
Associate Professor
Department of Agronomy
University of Agricultural Sciences
GKVK, Bengaluru-560 065, Karnataka, India

Manjanagouda Sannagoudar
Scientist (Agronomy)
Division of Grassland & Silvipasture Management
ICAR-IGFRI, Jhansi- 284 003, (UP) India

Gurunath Reddy
Sr. M.Sc. (Agri.) in Agronomy
Department of Agronomy
University of Agricultural Sciences
GKVK, Bengaluru-560 065, Karnataka, India

NIPA® GENX ELECTRONIC RESOURCES & SOLUTIONS P. LTD.
New Delhi-110 034

NIPA® GENX ELECTRONIC RESOURCES & SOLUTIONS P. LTD.

101,103, Vikas Surya Plaza, CU Block
L.S.C. Market, Pitam Pura, New Delhi-110 034
Ph : +91 11 27341616, 27341717, 27341718
E-mail: newindiapublishingagency@gmail.com
www: www.nipabooks.com

For customer assistance, please contact
Phone: + 91-11-27 34 17 17
Fax: + 91-11- 27 34 16 16
E-Mail: feedbacks@nipabooks.com

ISBN: 978-81-19103-69-0

Composed and Designed by NIPA®.

Foreword

The problem of global food security is one of the biggest challenges in front of Agricultural Scientists throughout the world. Food requirement of the ever-burgeoning population is alarming one and another side, horizontal expansion in area for cultivation crops is hardly or rarely impossible. At this juncture, in Indian Agricultural development, it is imperative to use resources in more scientific and rational way to increase the output /productivity per unit area per unit time over years without degrading the natural resources for future generations.

For better understanding of the fundamentals of agronomy clearly and easily, authors have presented the text book in a scientific and systematic manner. The present book titled **Fundamentals of Agronomy** suite to the need of students. I am happy that the authors have made painful efforts to bring this text book covering all Fundamental aspects of Agronomy.

I am confident that this text book will serve as reference book for students and teachers in the field of Agronomy. In particular, it will be a valuable study martial for undergraduate students.

Dr. B.S. Lalitha and Co-authors deserve commendation for their painful efforts and finally I congratulate them.

GKVK
Bengaluru-560 065

S. Rajendra Prasad
Vice Chancellor
University of Agricultural Sciences

Preface

An understating of fundamentals of agronomy is essential for scientific farming community to increase trend in crop production while ensuring sustainability of the system. Major problem in Indian Agriculture is low crop productivity but, population is growing at enormous rate. Therefore it is important to understand the fundamentals of agronomy to provide favourable environment for crop growth and development along with planning, programming and executing measures for efficient utilization of land, labour, capital and natural resources.

This text book is as per the syllabus prescribed by Fifth Dean's Committee on Higher Agricultural Education in India from the Academic year 2016-17. At present, several text books of Agronomy are available and this book is a further addition to that list with an added advantage of broad coverage of topics and well thought explanations of various fundamental aspects of Agronomy.

The authors warmly acknowledge their indebtedness to authors of books from which most of material has been freely drawn for compilation.

Finally authors are wish to acknowledge and express sincere thanks and gratitude to Vice Chancellor, University of Agricultural Sciences, Bengaluru, Dean (Agri.), College of Agriculture and Professor & Head, Department of Agronomy for providing the necessary encouragement to complete this work.

Authors

Contents

Foreword ... *v*
Preface ... *vii*

1. Agriculture as An Art, Science and Business ... 1
Agriculture ... 1
Agriculture as an art, science and business ... 2
Agriculture as a science ... 2
Agriculture as a business ... 3
Branches of agriculture ... 4

2. Agronomy and its Scope ... 7
Role of agronomist ... 7
Basic principles of agronomy ... 9

3. History of Agriculture Development in India and Karnataka ... 11
Agricultural systems of the world can be divided into 4 groups ... 11
The Agriculture history can be divided into 3 Phases ... 12
Theodore De Saussure (1804) concluded from his experiment that ... 14
Agricultural research / education in India ... 15

4. Factors Affecting Crop Production ... 19
External or environmental factors ... 20
High temperature stress ... 22
Low temperature stress ... 22

5. Importance and Scope of Agriculture ... 29

6. Classification of Crops 33

1. Based on range of cultivation 33
2. Place of origin 33
3. Botanical/taxonomical classification 34
4. Commercial classification 34
5. Agronomic classification 35
6. Seasonal classification 36
7. According to ontogeny 37
8. According to cultural requirement of crops 37
9. Based on textural suitability of soils 38
10. Based on tolerance to problem soils 38
11. According to cultivation of crops 38
12. According to the depth of root system 39
13. According to the tolerance to hazardous weather condition 39
14. According to water supply 39
15. According to method of sowing / planting 39
16. Based on length of duration of crops 40
17. Based on method of harvesting 40
18. According to post harvest requirement 40
19. Based on climatic condition 41
20. According to important Uses 41

7. Seeds and Sowing 43

Characteristics of a good seed 44
Factors influencing seed germination 45
Kinds of seed dormancy 46
Types of dormancy 46
Methods of dormancy breaking 46
Establishment of field crops may be grouped into 47

8. Soil and its Components, Fertility, Productivity and their Management 51

Soil is composed of 2 major components 51
Textural management of soil 53
Management 54
Factors affecting soil structure 54

Soil fertility and Productivity 58
Soil fertility evaluation 59

9. Tillage and Tilth 61

Objectives of tillage 61
Disadvantages of tillage 62
1. Primary tillage / Ploughing 64
2. Secondary tillage 65
3. Layout of seed bed 66
4. Tertiary tillage /leveling 66
Implements used for different operations 66
Zero tillage is practiced where 67
Advantages of zero tillage 68
Disadvantages of zero tillage 68
Factors influencing tillage 69

10. Crop Density and Geometry 71

Plant factors influencing crop density 71
Environmental factors influencing crop density 72
Plant density and yield relationships 72

11. Crop Nutrition-Manures, Fertilizers and Nutrient Use Efficiency 75

i. Based on quantity of nutrients present in plants are grouped into 76
ii. Based on their function in plants 76
iii. Based on mobility 77
Mechanism of nutrient movement in soil 78
Diagrammatic representation of occurrence of soil nutrient deficiency in India 79
Factors influencing composition of FYM 80
Concentrated organic manure 84
Based on the type of micro-organisms, bio-fertilizers are classified as 85
Based on the association, bio-fertilizers are grouped as 86
Fertilizers are classified into 86
General characteristics of nitrogen fertilizers 88
Phosphatic fertilizers 90

Calculation of Urea 95
Calculation of SSP 95
Calculation of MOP 96
Fertilizer especially secondary and micro-nutrients are also applied to the plant as: 98
Time of fertilizer application can be divided into 99

12. Growth and Development of Crops and Plant Ideotypes 101

Characteristics of growth 101
Types of growth rate 101
Plant growth regulators 102
Phytohormons 102
Classification of plant growth hormones (PGRs) 102
Maize 104
Cotton 104
Rainfed upland rice 104
Ideotype for Dryland Farming 105

13. Cropping Systems and its Principles 107

Benefits of cropping systems 112

14. Crop Adaptation and Distribution 113

Types of Ecosystem 113
Based on water requirement plants 115
Theories governing crop adoptation and distribution 116
World centres of origin of cultivated plants 117

15. Crop Management Technologies in Problematic Soils 119

1. Soil acidity 119
Causes for soil acidity 119
Management of acidic soils 120
2. Soil Alkalinity 120
Causes for alkalinity 120
Reclamation / management 121

Saline soils 121
Alkaline soils 121
3. Water logged / Marshy lands 121
Major soils of India 121

16. Harvesting and Threshing of Crops 125

Methods of harvesting 126
Threshing and shelling 126

17. Weeds: Importance and Classification 129

Plants are differentiated into 129
Importance of weeds 130
Losses caused by weeds 132
i. Based on life span or Ontogeny 135
ii. Based on ecological origin 137
iii. Based on soil type (Edaphic) 137
iv. Based on place of occurrence 137
v. Based on Origin 138
vi. Based on cotyledon character 138
vii. Based on soil pH 139
viii. Based on morphology 139
ix. Based on nature of stem 139
x. Based on specificity 140
xi. Classification according to association 141
xii. Based on association with human affairs 142
xiii. Noxious and objectionable weeds 142
Crop weed competition and crop weed interference 143
Critical period of crop weed competition 144
Critical period of weed competition for important crops 145
Factors affecting crop-weed competition 145
Crop Weed Interference 148
Concept of weed management-principles and methods 149
Concept in weed management 149
Principles of weed management 150
Methods of weed management 150
Physical or Mechanical weed control / management 152
Merits of physical or mechanical method 154
Demerits of mechanical method 155

Merits of cultural method 157
Demerits of cultural method 157
Qualities of bio-agent 158
Mode of action 158
Biological control agents 159
Insects 159
Merits of biological agents 160
Demerits of biological agents 160
Objective and scope of herbicide usage 161
Merits of chemical weed control 162
Demerits of chemical weed control 162
Classification of herbicides 163
1. Based on Placement 163
2. Based on Selectivity 163
3 Based on Mode of action / translocation 163
4. Based on Time of application 163
5. Based on the persistence / residual toxicity 164
Based on chemical structure 165
Need for preparing herbicide formulation 165
Types of formulation 165
Methods of application 166
Soil application of herbicides 167
Foliar Application 168
Methods of treating brush and trees 168
i. Difference in absorption 170
ii. Differential translocation 171
iii. Differential protoplasmic resistance 171
iv. Differential rate of deactivation 171
v. Differential carbon metabolism 172
Precautions in storage and handling of herbicides 172
Integrated weed management (IWM) 173
Need for integrated weed management 173
Advantages of integrated weed manaement 174
Parameters to study weed control effect 174
Allelopathic effect of weeds on crops 176
1. Maize 176
2. Sorghum 176

3. Wheat 176
4. Sunflower 176
Allelopathic effect of crop plants on weeds 176
Allelopathic effect of weeds on weeds 176

References 179

Annexure 181

National Institutions for Agricultural Research 181
International Institutions for Agricultural Research 183
Year of Establishment of Institutes 184

Chapter 1

Agriculture as An Art, Science and Business

Agriculture

The term agriculture is derived from Latin word ***ager*** or ***agri*** meaning **soil** and ***cultura*** meaning **cultivation**. Agriculture is a broad term encompassing all aspects of crop production, livestock farming, fisheries, forestry *etc*. Agriculture is a branch of applied science and is the art of farming including the of cultivating the soil for producing crops and raising livestock. There are three main spheres of agriculture-geoponic meaning cultivation in soil, hydroponic meaning cultivation in water and aeroponic meaning cultivation in air.

Agriculture is a productive unit where the gifts of nature like land, light, water and temperature are integrated into a single primary unit that is crop plant which is indispensable for human beings. The secondary productive units of agriculture are animals including livestock, birds and insects which feed on the primary units and provide concentrated products such as meat, milk, wool, eggs, honey, silk and lac.

The important cultural energies utilized for the production and protection of agricultural commodities are :

Natural Resources

1. Land, water, light and other resources of environment.

Added Resources

1. Irrigation and drainage
2. Organic, biological and chemicals fertilizers
3. Farm equipment's and draft power.

The word Agriculture may be expanded as ***Activities*** on the ***Ground*** for ***Raising Intended Crops*** for ***Uplifting Livelihood Through*** the ***Use*** of ***Rechargeable Energies***.

Agriculture as an art, science and business

Agriculture as an art embraces knowledge of way of performing the operations of the farm in skillful manner, but does not necessarily include an understanding of the principles underlying farm practices. The skill may be physical and mental. Some farmers are able to do farm operations quickly and more efficiently than others with great ease. They are said to be skillful in their art. The ability and capacity to plough well, to make a good hay stack, to handle animals properly, to handle the farm implements and proper sowing of seed, manures, fertilizers, *etc.*, are examples of physical skills which are obtained through experience. Similarly, deciding the proper time for agricultural operations like ploughing, selection of cropping pattern to suit the soil and climate, adopting improved agricultural practices, time of production to get better prices are examples of mental skills involved in farming. Both mental and physical skills are essential for successful farming.

Agriculture as a science

Scientific principles were used in agriculture for increasing production and quality of crops and in their ultimate disposal to earn better price and income. Recent advances made in Genetics and Plant breeding and developed many varieties suit to various soils, climate and consumer tastes, use of fertilizers, pesticides, fungicides, herbicides and growth regulators, introduction of labour saving machineries, *etc.*, have completely revolutionized agriculture for increased crop production.

In Plant Breeding and Genetics, strains were evolved to suit different conditions of soils, climate and consumers tastes. Photo insensitive and fertility responsive high yielding strains available in rice and wheat have induced green revolution in different parts of world. Research findings in isolating genes responsible for quality protein have helped in screening varieties rich in lysine, methionine and tryptophan.

Agriculture as a business

Agriculture in developed countries is a business driven process. The farmers in these countries cultivate the crops to earn higher income with less investment and agriculture is practiced on a business line and land is considered as workshop or factory to generate returns. Traditional farming is handed over from father to son without any effort to increase production. Mechanization and commercialization have been introduced in these countries to get maximum profit. In developing and under developed countries, agriculture is considered as a way of life and practicing on a subsistence level. An industry run business motive will work more efficiently and diligently with the object of getting more profit from fewer investments. Similarly in agriculture the land is workshop from where greater returns are expected. To obtain these returns, the factors like land, labour and capital are utilized in such a way that greater profits are obtained. Just like any industrial business, the farming operations also compromise such problems as production, consumption, barter, trade, employment and employees, marketing, transportation, internal trade, economic interdependence, taxation, customs and tariff in their relation between the farmer and other persons such as merchants, manufactures, tax collectors and transporters. In order to tide over these factors and to earn greater profits, agriculture is mechanized and commercialized to run a business starting from selection of an enterprise for one's livelihood, agriculture is treated as any other business in the developed countries. After the decision has been taken for farming, the choice of growing a particular crop in particular area is decided on the possibility of earning the maximum profit.

Difference between Agriculture and Industry

S.No.	Agriculture	Industry
1.	Biological process in which plants convert solar energy to chemical energy	Mechanical process
2.	Primary products of agriculture is organic in nature	Majorly inorganic
3.	Primary resources are plant and environment	Depends on agriculture for raw material
4.	Subjected to natural calamities and production is not under control	Can be protected from natural calamities and production can be controlled.

Branches of agriculture

Major branches: Agronomy, Genetics & Plant Breeding, Soil Science & Agricultural Chemistry, Agricultural Entomology and Plant Pathology.

Allied branches: Horticulture, Forestry, Animal Science, Agril. Microbiology, Fishery Science, Crop Physiology, Plant Biotechnology, Agricultural Engineering, Agril. Economics, Agril. Statistics and Home Science.

1. **Agronomy:** It deals with the production management of various crops which includes food crops, fodder crops, fiber crops, sugar, oil seeds, *etc.*

2. **Horticulture:** Branch of agriculture deals with the production of Flowers, Fruits, Vegetables, Ornamental plants, Spices, Condiments and Beverages.

3. **Forestry:** It deals with large-scale cultivation of perennial trees for supplying wood, timber, *etc.,* and also raw materials for industries.

4. **Animal Husbandry:** Animal Husbandry is the rearing of various animals for meat, milk, wool, draft energy *etc.*

5. **Fishery Science:** It is for marine fish and inland fishes including shrimps and prawns.
6. **Agricultural Engineering:** It is an important component for crop production and horticulture particularly to provide tools and implements. It is aiming to produce modified tools to facilitate proper animal husbandry and crop production.
7. **Home Science:** It deals with the effective conversion of raw produce into consumable with value addition.

Chapter 2

Agronomy and its Scope

Agronomy is a branch of applied science deals with the principles and practices of field crops production and management of soil for higher productivity. The term Agronomy is derived from Greek word ***agros*** meaning **field** and ***nomos*** meaning **to manage**.

Norman (1980) defined Agronomy as the science of manipulating the crop environmental complex with dual aims of improving Agricultural productivity and gaining a degree of understanding of the process involved.

In recent times, Agronomy has assumed newer dimensions and can be defined as **'A branch of Agricultural science that deals with the methods which provide favourable environment to the crops for higher productivity'.**

Role of Agronomist

Agronomist exploits the knowledge of the basic and applied sciences to maximize crop production at minimum cost. Agronomists are concerned with deriving efficient technology of crop production activities like.

1. **Land Preparation:** Developing efficient and economic field preparation methods depending on crop, season and resource availability. Developing machineries in co-ordination with Agricultural Engineers.
2. **Selection of crops:** Involved in selecting suitable crop and variety in accordance with the season, soil and farmers preference.

3. **Sowing / Planting:** Develop efficient method of sowing / planting and standardized spacing / plant geometry to maintain optimum plant population and in turn better yield.
4. **Nutrient management:** Identify various nutrients required by the crop taking crop demand and soil fertility in to consideration. He is also responsible for deriving optimum time and method of application of these nutrients.
5. **Weed management:** Agronomist is responsible for selecting better weed management practices *viz.,* mechanical / physical or cultural / agronomic or biological or chemical or integration of two or more methods.
6. **Irrigation management:** Selection of proper irrigation method, time and quantity of application based on crop water need and its availability.
7. **Crop planning / cropping system:** Identifying the crop sequence, inter / mixed cropping for efficient utilization of available resources to maximize farm income.
8. **After care:** Management of crop after planning/sowing till harvesting including the schedule of plant protection measures, inter-cultivation etc.
9. **Harvesting and Processing:** Responsible for deriving efficient method and time of harvesting and processing.
10. **Decision making:** On marketing / sale of produce (Time and place of marketing). Overall, **Agronomist** is like a **General physician.**

Agronomy is a dynamic discipline, Agricultural practices are modified or new practices developed with the advancement of knowledge and better understanding of plant and environment. For example, availability of chemical fertilizers has necessitated the generation of knowledge on the method, quantity and time of application of fertilizers. Similarly, availability of herbicides has led to development of vast knowledge about selectivity, time and method of application of herbicides. Gigantic irrigation projects are constructed to provide irrigation facilities. However, these projects created side effects like water logging and salinity. To overcome these problems, appropriate water management practices are developed.

Population pressure is increasing but the area under cultivation is static. As such, more number of crops must be grown on the same piece of land in a year. As a result intensive cropping has come into vogue.

Similarly no tillage practices have come in place of clean cultivation as a result of increase in cost of energy. Likewise, new technology has to be developed to overcome the effect of moisture stress under dry land conditions. As new varieties of crops with high yield potential are available then need to develop package of practice to exploit their full potential.

Factors restricting increased agricultural production are low soil fertility, crop varieties of low genetic yield potential, inadequate control of diseases and insects, non availability of production inputs, government economic policies affecting agriculture and weak research and extension programmers.

Restoration of soil fertility, preparation of good seed bed, use of proper seed rates, correct dates of sowing, proper conservation and management of soil moisture and proper control of weeds are agronomic practices to make our finite land and water resources more productive. Thus agronomy discipline has great scope.

Basic Principles of Agronomy

Agronomic principles are the ways and means for better management of soil and environment for profitable crop production. **Basic principles of Agronomy include:**

1. Planning, programming and executing measures for efficient utilization of land, labour, capital and natural resources for profitable crop production.
2. Proper field management by way of attending to land leveling for checking soil erosion, provision of field channels for irrigation

and drainage, watershed approaches for water and soil conservation, *etc.*

3. Planning for appropriate cropping systems, include multiple cropping and mixed cropping or intercropping to ensure harvest even under adverse environmental conditions. Planning for farming systems minimizes risk in farming.
4. Choice of crops and varieties adaptable to a particular agro climate, land situation, soil fertility, season and methods of cultivation for remunerative farming.
5. Optimum crop stand establishment through ideal seedbed and quality seed for efficient utilization of applied inputs and natural resources.
6. Timely application of manures and fertilizers and emphasis on integrated nutrient management systems for soil fertility and productivity improvement.
7. Proper irrigation scheduling for efficient use of scarce irrigation water and applied inputs and planning for drainage at times of high intensity rainfall.
8. Adoption of management practices including intercultural operations to realize maximum benefit for monetary and non-monetary inputs. Adequate input supply should be a substitute for deficiencies in management practices.
9. Adoption of timely need based timely plant protection measures against weeds, insects, pests, pathogens as well as climate hazards.
10. Planning for timely harvesting of crops to reduce harvest losses and to release land for timely planning of succeeding crops. Adoption of suitable post harvest technologies.

Chapter 3

History of Agriculture Development in India and Karnataka

It is supposed that man was evolved on earth about 15 lakh years ago. This man started to move by standing erect on his feet and called *Homo erectus* L. Later the modern man was evolved called *Homo sapiens* L. (Sapiens means learning habit) due to his continuous learning habit. It is difficult to trace the exact period of beginning of agriculture. Archeological evidences indicated that, Agriculture started about 50,000 years ago.

Hunting was the primary source of food in olden days and was prevailed for long time. They obtained his food through domestication of some animals like dogs, horse, cow, buffalo etc. They live in the periphery of the forest and migrate from one place to another in search of food. It was not comfortable and they started enjoying by settling at one place near the river and started agricultural system.

Agricultural Systems of the world can be divided into 4 groups

1. Shifting cultivation
2. Simple sedentary cultivation
3. Simple plough cultivation
4. Advanced / modern agriculture

1. **Shifting cultivation:** Moving from one place to other specially the river beds by clearing forest through fire and broadcasting the crop seeds like paddy, chilli, redgram *etc.,* it is also called as Jhum Cultivation.

2. **Simple sedentary cultivation:** People use to dig pits / holes with sharp spade or pickaxe and plant / sow seeds. It is common in Asian countries and crops cultivated are banana, coconut, areca nut, root tubers *etc.*
3. **Simple plough cultivation:** Practiced with development of small local ploughs and animal draft power. Ex: Rice cultivation.
4. **Advanced / modern agriculture:** It is based on modern machineries and adoption of technologies developed over time.

The Agriculture history can be divided into 3 Phases

1. Ancient period (From 800 BC to the fall of Rome-476 AD)
2. Medieval period (From 476 AD to 1600 AD)
3. Modern period (From 1600 AD onwards)

1. Ancient period

Period	Event
Before 10,000 BC	Hunting and gathering of fruits, grains in forest area and animals
7,500 BC	Shifting cultivation of crops- Wheat and Barley
3,400 BC	Invented wheel (out of stone)
3,000 BC	Bronze used for making tolls
2,900 BC	Invented plough, started irrigated farming
2,300 BC	Cultivation of chickpea, cotton and mustard
2,200 BC	Rice cultivation
1,500 BC	Sugarcane cultivation
1,400 BC	Used iron for making tools
1,000 BC	Use of iron plough
430 to 330 BC	Xenophan-Greek historian mentioned about ploughing of green material to enrich soils
234 to 149 BC	Cato gave importance to ploughing and manuring
70 to 19 BC	Virgil mention about crop rotation and inclusion of legumes to enrich soil fertility.

Mentions about agriculture have also been made in Vedas (Rigveda, Yajurveda, Atharvanaveda and Samaveda).

Rigveda: Green cultivation, cattle rearing, use of ploughs & sickles.

Atharvanaveda: Different crops grown in different season.

2. **Medieval period:** During the period of Buddisht's, Gupta's, Maurya's and Magadha's, mention has been made on agriculture. For instance:
   a) A Greek ambassador **Megasthanis** came to India and mentioned how Indian people giving importance to maintain soil fertility and crop production.
   b) **Kautilya**, a famous economist mentioned the role of a king in protecting agriculture. He emphasized for drying seeds/ grains and their storage for future use.
   c) **Saint Manu**, in his 'Manusmrithis' has mentioned about the punishment given against implement theft.
   d) **Saint Parashara (1000 AD)** in his book 'Krishiparashara' mention the operations for paddy for different season. He emphasized manuring, ploughing, puddling, thinning, weeding and irrigation.
   e) **Peter Decresenzi (1230-1307 AD)** collected many literatures related to Agronomy in his book 'Opus Ruralium Kamo Daram'. Hence, he is considered as **Father of Agronomy.**
3. **Modern period / Development of scientific agriculture**

Scientific agriculture was started during 16th century. Experimentation technique was started by **Francis Bacon (1561-1624 AD)**. He concluded from his study that water is the principle element for plant.

Van Helmont (1572-1644 AD) conducted a pot experiment on willow tree, popularly known as **'Willow Tree Experiment'**. He took a willow plant of weight 5 Pounds and planted in a pot containing 200 pounds soil. The plant was monitored for 5 year by continuous watering. By the end of 5th year, the willow tree weighed 169 Pounds and the weight of soil was 198 Pounds. He concluded that water is the slow requirement for plants.

Jethrotull (1674-1741) mentioned soil particles are essential for plant growth. He developed **Horse hoe for tillage**. He suggested providing more soil particles to plant by tilth and addition of animal excreta and manures to soil, which make it friendlier for plant growth. He invented **'seed drill'** in 1731 for row sowing. He also published a book 'Horse hoeing husbandry' in 1733. He is considered as **Father of Tillage.**

Arthur Young (1741-1820) published **Annals of Agriculture** literature on agriculture.

Priestly (1780) mentioned that plant take CO_2 from carbonic acid of soil and give up O_2.

Theodore De Saussure (1804) concluded from his experiment that

1. Plants absorb CO_2 from atmosphere and not from soil.
2. 'N' essential for plant growth taken from soil.
3. Plant roots are selective in absorption and absorb only water from soil leaving salts. Hence, salts accumulate in soil.
4. Mineral matters are not synthesized in the plants but are absorbed by roots from the soil.

The large scale field experimentation was started by **J B Boussingault (1834)**. He is recognized as **Father of Field Experimentation** and worked on mineral nutrition of plants.

Justas Van Liebig, a German scientist considered as **'Father of Agricultural Chemistry'** opinion that, growth of plant was proportional to the amount of mineral substance available in the soil. He proposed a law called **Liebig Law of Minimum.** It states that 'The growth of plant is limited by the plant nutrient which is present in smaller quantity while all other being adequate'.

JB Lawes and JH Gilbert (1843), started an agricultural experiment station to conduct long term fertilizer trial and to test the superphosphate they developed by the action of rock phosphate and H_2SO_4 at Rothamsted.

George John Mendel (1856) discovered laws of heredity.

Charles Darwin (1876) proposed theory of evolution and inheritance of acquired character.

PMA Millardet (1880) discovered 'Bordeaux Mixture' ($CuSO_4$ + $Ca(OH)_2$ + H_2O :: 1:1:100 ratio) to control downy mildew in grapes.

Thomas Malthus (1898) proposed theory of population growth, which states "Population in India is growing in geometrical and food grain in arithmetical progression. Human would run-out of food in spite of advances in agriculture due to limited land and yield potential of crops".

Blackman (1905) Proposed **Theory of Optima and Limiting Factor**. Which states that 'When a process is conditioned as to its rapidity by a number of separate factors, the rate of the process is limited by the pace of slowest factor'.

Mitscherlich (1909) proposed **Law of Diminishing Returns**, which states Increase in growth with each successive addition of the limiting element is progressively smaller and the response is curvilinear.

Wilcox (1929) proposed **Inverse Yield Nitrogen Law**. It states that the growth or yielding ability of any crop plant is inversely proportional to the mean nitrogen content in plant. He also published a book called **Wilcox Agro-Biology**.

Agricultural Research / Education in India

No Agricultural experiment stations were initiated in India till 1903. There was famine in 1877, 1889, 1892, 1897 and 1900 in India led to decreased population due to scarcity of food.

Lord Curzon (1898-1905) constructed great canal in Western Punjab. Famine commission appointed in 1888, 1898 and 1901 stressed the importance of agricultural research. As a consequence, he started **Imperial Agricultural Research Institute** in 1905 at Pusa, Bihar. His period is called as **'Golden Period of Agriculture'.** During his regime, Agricultural Colleges were started at Poona, Dehradun, Saidapet (Coimbatore) Province. Also **Hebbal Farm** at Bangalore in Mysore Province was started in 1899. Later it was upgraded as Agricultural school in 1912-13 to offer Diploma in Agriculture.

Due to earthquake in 1934 at Pusa, Bihar, Imperial Agricultural Research Institute was shifted to New Delhi in 1936. Later, it was renamed as Indian Agricultural Research Institute (IARI). Which is one of the largest research institutes in the country and also taking Post-Graduate education.

Further, based on the 1928 Royal commission recommendation, **Imperial Council of Agricultural Research was** started at **New Delhi in 1929**. Which was later changed as Indian Council of Agricultural Research (ICAR) with 3 main objectives :

1. To Co-ordinate the education in agriculture, animal husbandry, fishery, research and its application (Extension)
2. To finance various research institutes, Agril. Universities, schemes/ projects all over the country
3. Disseminate the knowledge gained by research through literature.

ICAR is the sole body which controls all Agricultural Research Institutes in India. Several research institutes both at National and International level were set up to take location specific research.

After independence, ICAR adopted to Land Grant Colleges. The first Land grant College was started at Panth Nagar (Uttaranchal) in 1962. University of Agricultural Sciences, Bangalore was established in 1965 with Three objectives *viz.,* Teaching, Research and Extension.

There are 67 Universities under ICAR with research institutes on its own. Presently, in Karnataka, 5 Universities are coming under ICAR.

Deemed Universities-4

Institutions-64

NRC-15

National Bureaus-6

Project Directorates-13.

Chapter 4

Factors Affecting Crop Production

Higher plants demand certain basic things for their growth. As many as 52 factors (basic things) influencing crop growth have been identified :

- 100 per cent of the land has sufficient CO_2 in the atmosphere and sunlight
- 83 per cent have favourable temperature
- 64 per cent have favourable topography
- 46 per cent have reliable rainfall
- 40 per cent have satisfactory fertile land

But only 7 per cent have combination of these above factors to make feasible for crop production. Without advanced technology and management this also declining still in future.

The factors influencing crop growth / production can be broadly grouped into 2 groups :

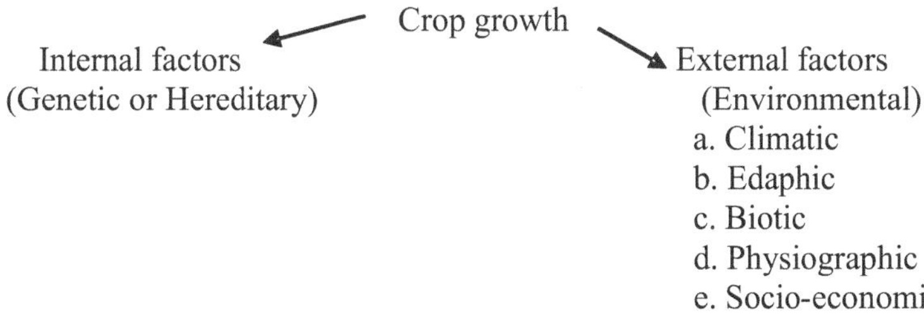

Genetic factors: The increase in crop yields and other desirable characters are related to genetic makeup of plants. Breeders and Bio-technologists have incorporated many desirable characters and developed HYV and hybrids. The characters considered are :

1. High yielding ability
2. Early maturity
3. Resistance to lodging (Strengthening of stalk)
4. Tolerance to drought, flood and salinity
5. Tolerance to insect pests and diseases
6. Chemical composition of grains (high percentage of oil, protein *etc.*)
7. Quality of grains (fineness, coarseness, *etc.*)
8. Quality of straw (sweetness, juiciness, *etc.*)

These characters are inherited and transferred from one generation to another through genes.

External or Environmental Factors

Crop growth and development is primarily governed by environmental conditions. Success or failure of crop is related to the weather condition that prevails during crop growth. Control of environment under field condition is more difficult, costly and often impossible. Therefore, it is essential to study these factors in order to cope up with them for better crop growth.

Climatic Factors: Climate is a generalized weather or summation of weather conditions over a given region during comparatively longer period. Where as weather is a state or condition of atmosphere at a given place and at a given time.

The important climatic elements that affect crop production are:

1. Precipitation
2. Temperature
3. Atmospheric humidity
4. Solar radiation
5. Wind
6. Atmospheric gases

1. Precipitation: Precipitation includes all forms of water that fall on earth from the atmosphere such as rainfall, snow, hail, dew, fog *etc.* It is the source of water for crop production.

Rainfall influences the vegetation of a place, precipitation amount and distribution greatly affects the choice of crops and cultivars, low and uneven distribution of rainfall (dry lands) compel the farmers to go for drought resistant crops like pearl millet, sorghum and minor millets. Desert climate forces to go for grasses and shrubs. Distribution of rainfall is most important than total rainfall for successful crop production.

Deficient rainfall leads to desiccation of plant, reduce yield and result in plant death.

About, 55 per cent of the world land receives	< 51 cm rainfall	Semi arid
20 per cent	51 – 102 cm RF	Sub humid
11 per cent	102 – 152 cm RF	Humid
14 per cent	> 152 cm RF	Super humid

Depending on the water need, the plants are grouped into 3 groups

1. **Xerophytes:** Plants survive under very low moisture condition and withstand drought even for prolonged period. They are also called as Desert plants. Ex. Agave.
2. **Mesophytes:** They require normal supply of moisture for their growth and grow well when soil is moist. Ex. Wheat, Maize, Pulses.

3. **Hydrophytes:** They are water loving plants, grow well only when the soil is wet or full of water. Ex. Rice, Aquatic plants.

2. Temperature: Temperature is a measure of intensity of heat energy or the degree of hotness/coolness. Physiological process with in the plants is controlled by air temperature. The growth increases with an increase in temperature to a certain extent but reduces after a limit. The range of temperature for optimum growth of most crops is between 15 to 40^{0}C. Temperature influences distribution of vegetation including crops.

Extreme high and low temperature causes stress in plants

High temperature stress

1. Causes reduction in absorption and assimilation of nutrients.
2. Affect shoot growth.
3. Abortion of pollens – reduce viability and cause female sterility.

Low temperature stress

1. Survival, cell division and elongation is strongly affected.
2. Absorption and translocation of water is affected due to low permeability of water.
3. Reduced translocation and gas exchange resulted in poor photosynthetic rate and in turn the crop yield.

3. Atmospheric humidity: Humidity refers to the water vapour suspended in the atmosphere. Relative Humidity (RH) influences the crop water requirement. RH of 60-70 per cent appears optimum for most crops. More RH leads to outbreak of pest and diseases.

4. Solar radiation: Solar radiation is the source of energy for all physical processes taking place in the atmosphere. Growth and development of crops is a function of solar radiation. Solar radiation dictates the rate of photosynthetic process. All physical process like photosynthesis, chlorophyll synthesis, evaporation, heating of soil and air to regulate temperature besides germination, leaf expansion, stem growth, flowering, fruiting *etc.,* taking place in the soil, plant and environment are depending on solar radiation.

Amount of radiation (1.94 cal cm^{-2} min^{-1}) is emitted continuously from the sun. The amount of solar energy received per unit area of surface held at right angle to the sun rays at the top of the atmosphere is called solar constant.

Light affect plants growth through its intensity, duration and quality:

1. **Light intensity:** The total quantity of light received in unit time can be considered as light intensity, which is expressed in Foot Candles / Lux. The optimum light intensity for most of the crops is around 1700 Lux.

There are certain plants which grow luxuriously with high light intensity, such plants are called **Heliophytes** (Sun-loving) Ex. Sunflower. The plants growing under low light intensity / shade condition are called **Sciophytes** (Shade loving) Ex. Coffee.

The light intensity at which the photosynthates produced through photosynthesis is completely made use for its respiration is called **Light saturation point.** It varies from sciophytes (27 Lux) to heliophytes (4200 Lux). Higher light intensity may disintegrate chlorophyll.

2. **Quality of light:** The process of photosynthesis in green plants depends on quality of light. Radiation up to 250 nm in UV spectrum is harmful to most of the plants. Though, radiation from 300 nm is photosynthetically active, visible spectrum (400-750 nm) is more effective. Visible spectrum composed with seven colour range VIBGYOR.

Violet - 400 to 435 ηm　　**Green** - 490 to 574 ηm

Orange - 595 to 626 ηm　　**Indigo** - 435 to 460 ηm

Yellow - 574 to 595 ηm　　**Red** - 626 to 750 ηm

Blue - 460 to 490 ηm

Within the visible range, the principle wavelength absorbed in photosynthesis are Violet to Blue (400-490 nm) and Orange to Red (595-750 nm). Red light is found most favourable. Radiation above 750 nm (Infra Red radiation) is harmful to photosynthesis.

3. **Duration of light:** Light intensity and duration determine the amount of light that a plant receives. The length of the day is more important than intensity. The response (Flowering) of the plants to the relative length of day and night is known as **photoperiodism**. Different species and in some cases different cultivars of the same species responds differently to the length of day and night. The photoperiod required to induce flowering is called **critical day length**.

Based on the photoperiodic response, the plants are grouped as:

1. **Long day plants:** Crops that reproduce (Flower) normally when the light period is longer than 12 hours. Ex. Rabi Crops-Wheat, Barley, Sugar beet, Potato *etc.*
2. **Short day plants:** Crops that reproduce (Flower) normally when the light period is shorter than critical minimum (<12 hours). Ex. Tobacco, soybean, millets, Rice *etc.*
3. **Day neutral plants:** Plants that are non-responsive to relative length of day & night. Ex. Tomato, Sunflower, Buck wheat *etc.*

5. **Wind:** Wind is the air in horizontal motion which travels from high pressure areas to low pressure areas. The basic function of wind is to carry moisture and heat and supplies fresh CO_2 for photosynthesis. Mild wind (4-6 km hr^{-1}) is essential for crop production. High wind velocity cause mechanical damage to the crops like banana, sugarcane, orchards crops, *etc.* Wind is necessary for pollination and seed formation. Wind increases the evapotranspiration and responsible for spread of pests and diseases. Heavy wind leads to soil erosion. Wind also responsible for rainfall, mild wind encourages pollination and gaseous exchange.

6. **Atmospheric gasses:** Atmospheric air is composed of 78.09 % Nitrogen, 20-98 % Oxygen, 0.93 % Argon & 0.03 % CO_2. CO_2 is most important for photosynthesis and taken by the plant through diffusion in stomata. Oxygen is important in respiration of both plants and animals. Nitrogen is important plant nutrient and Atmospheric nitrogen is fixed in the soil by lightening, rainfall & microbes. Certain gasses like SO_2, CO, CH_4, HF, CFC released from petrochemicals are toxic.

Edaphic factors: Are the soil factors and soil provides physical anchorage to plants and acts as a store house of water, nutrient and air.

The soil factors that affect crop growth are:

1. Soil moisture
2. Soil temperature
3. Soil organic and mineral matter
4. Soil organisms
5. Soil air
6. Soil reaction

1. **Soil moisture:** More than 90 per cent of plant tissue consists of water which is absorbed from soil through roots. It is held by soil particles by cohesive and adhesive forces. Rainfall, irrigation and drainage govern the soil moisture. Soil moisture is the medium for absorption and translocation of mineral nutrients. Soil moisture maintains soil temperature, regulates soil aeration and together these dictate the activity of biotic factors in soil.

   Soil texture, structure and depth determine its moisture retention besides the organic matter. Efficient use of inputs in crop production depends on optimum level of soil moisture.

2. **Soil air:** Composition of soil air differs from that of atmosphere. Soil air has 10-100 times higher CO_2 than atmosphere. Aerobic micro organisms, sprouting seeds, root respiration demand ample supply of $Oxygen_2$ in the soil. Rice can withstand anaerobic condition because of the presence of aerenchyma tissue. Soil moisture content / irrigation dictate the soil air and affects root respiration.

3. **Soil temperature:** Soil temperature influence root growth and functions. Soil temperature is particularly important during crop germination and establishment. Availability and absorption of nutrients and water maximum absorption occurs at 20-30^{0}C. Regulate the microbial activity. Root growth and development especially in edible roots such as turmeric, tapioca, sugar beet, potato, groundnut *etc.*
4. **Soil organic matter:** Soil organic matter is derived from death and decaying plant parts / animals and their external addition. Soil organic matter supplies major and minor nutrients. Improve soil structure and aeration. Improves water holding capacity & moisture availability. Source of food for micro-organisms.
5. **Soil organisms:** The organic matter in the soil is decomposed by different micro-organism to release nutrients. The different kind of organisms are :

   Macro-flora: Roots of higher plants.

   Macro-fauna: Earthworms, grubs.

   Micro-flora: Algae, Bacteria, fungi, Actinomycetes.

   Micro-fauna: Protozoa, Nematode, Mites, insects etc.

There are both beneficial and harmful organisms in the soil and their activity in turn depend on soil moisture, temperature and aeration.

Beneficial organisms like Azatobacter, Azospirullum, Rhizobium, BGA, fix atmospheric nitrogen, phospho bacteria release the fixed 'phosphorus' to the plants.

The harmful organisms are soil borne insect, pathogen and weeds.

6. **Soil reaction / pH:** Availability of plant nutrient depends on soil reaction. pH is the negative logarithm of hydrogen ion concentration. Soil acidity and alkalinity have profound influence on crop productivity. Most of the plant nutrients are freely available at neutral pH. Acid soil with higher Fe, Al & Mn reduce the availability of P, availability of Mo, Fe, Al, Mn will

reduce with alkalinity. Also, pH influence the microbial activity and water absorption.

Biotic factors: Includes the beneficial / harmful effects caused by other plants, animals, micro-organisms on crop plant.

Plants may be complementary / competitive among each other. Competition between plants occurs when the supply of nutrients, moisture and sunlight is limited. This competition may be between same crops or between different crops or weed & crop.

When different crop of cereal & legume grown together, mutual benefits among them results in higher yield (synergetic effect).

Animals: Soil fauna like Protozoa, Earthworm *etc.,* improve organic matter decomposition. Honeybees & wasps help in cross pollination. Large animals like cattle, goat *etc.,* may damage crop but add organic matter.

Physiographic factors: The nature of earth surface (slope / flat) known as topography influence crop growth directly. Steepness of slope dictates runoff and soil erosion. Increase in altitude cause a decrease in temperature and increase precipitation & wind velocity. Physiography dictates exposure of crop / soil to light & wind.

Socio-economic factors: Crop and cultivar selection determines the ultimate object of crop production. Scientific interventions lead to development of high yielding cultivars resistant to pest and diseases. Economic condition of the farmer greatly decides the input / resource use and in turn crop growth.

Chapter 5

Importance & Scope of Agriculture

Agriculture plays a key role in the overall economic and social well being of the country. Though the share of Agriculture in both GDP and employment has declined overtime, Agriculture still forms the backbone and occupies a pride of place :

1. Agriculture contributes 24 per cent to the Gross Domestic Product. GDP was declined from 39 per cent in 1983 to 25 per cent during 1999-2000 and now it 17.32 per cent.
2. Provides livelihood support for more than half of the population in the country. The sector provides employment to 56.7 per cent of the India's work force.
3. Agriculture has been the source of raw material for industries. Majority of the industries depending on agriculture for their raw materials. Ex: Textile (Cotton, Jute), Sugar (Sugarcane, Sugar beet), Small & medium scale industries like soaps, dyes, medicines, vitamins, preservation of fruits and vegetables, dhal milling, rice husking, Jaggery making, oil crushing, handlooms *etc.*
4. It provides large part of market for industrial goods *viz.*, seeds, fertilizers, pesticides, implements, machines like pump sets, trucks, tractors, power tillers, sprayers *etc.*
5. Support roadways, railways, airways and shipways for transport of inputs & produce.
6. Provides a large portion of India's export. Agriculture provides 12-15 per cent of national exports. The commodities exported are tea, coffee, sugar, cashewnut, spices, tobacco, cotton *etc.*

7. Agriculture maintain food security and in turn national security. Satisfactory agriculture production brings peace, prosperity, harmony, health and wealth.
8. "A man without food for 3 days will quarrel, for a week will fight and for a month will die".
9. Agriculture and allied sectors like Horticulture, Animal husbandry, Dairy & fishery are important in improving economic conditions, health and nutrition of human being.

Agricultural growth can be witnessed by increased food grains from 50 to 215 m t productivity from 522 to 1500 kg ha^{-1}, per capita food availability from 395 to 510 g day^{-1}, cotton from 3 to 13.5 m. bales, sugarcane from 57 to 330 m t eggs from 2 to 28 billion, vegetables from 20 to 90 m t.

This has happened mainly because of revolutions with development of technology. The major revolutions are :

Green Revolution	-	Food grains (50 to 272 m t)
White Revolution	-	Milk (17 to 70 m t)
Blue Revolution	-	Fish (0.75 to 5.0 m t)
Yellow Revolution	-	Oilseeds (5 to 25 m t)

Geometrical growth of population resulted in 1.2 billion in India. However, food grains production is moving arithmetically. The food grains demand by 2025 is estimated at 335 million tons over different years and 2017-18 food production was 272 m t.

In future, agricultural development in India would be guided not only by the compulsion of improving food and nutritional security, but also by the concerns for environmental protection, sustainability and profitability.

We need to achieve the food grains production growth rate by 2.7 per cent to meet the population growth rate of 2.2 per cent. Which is possible through :

1. Increasing the area under HYV and Hybrids
2. Adoption of improved production technology
3. Increasing area under irrigation
4. Efficient and integrated management of manures and fertilizers
5. Adoption of suitable crop / cropping system and formulation of cropping zones
6. Attentions to develop integrated pest management timely to avoid losses
7. Rain water management-efficient conservation & harvest followed by judicious use
8. Transfer of technology from lab to land efficiently and timely.

Chapter 6

Classification of Crops

Crops: Is a cultivated plant that is grown on large scale commercially, especially cereal, pulses, fruits or vegetables.

Classification of crops : Several crop plants are alike with respect to ontogeny (Development), morphology, anatomy, physiology and requirement of environment. Classification is done to generalize similar crop plants as a class for attaining better understanding of them. Field crops are classified in several ways.

1. Based on range of cultivation

a) **Garden crop:** Grown on a small scale in gardens. Ex. Kitchen garden, vegetable garden, flower garden *etc.*

b) **Plantation crop:** Grown on a large scale in estates and perennial in nature. Harvesting continues for a prolonged period after single planting. Ex. Tea, Coffee, Cacao, *etc.*

c) **Field crop:** Grown on a vast scale under field condition. They are mostly seasonal such as rice, wheat, cotton *etc.* Agronomy deals with field crops only.

2. Place of origin

a) **Native Crops:** Crops grown within the geographical limits of their origin. Ex. rice, barely, blackgram, green gram, mustard, castor, sugarcane and cotton, are native to India.

b) **Exotic or Introduced Crops:** Crops introduced from other countries, such as Tobacco, Potato, jute, maize, apple, *etc.* are introduced from other countries to India.

3. Botanical/taxonomical classification

Crop plants are grouped as orders and families in a systematic arrangement. This classification has certain advantage in understanding of morphological characters of crop plants with in the group :

a) Poaceae (Graminae): Ex. Cereals, millets and grasses.

b) Papilionaceae (Leguminaceae): Pulses, groundnut, berseem, *etc.*

c) Cruciferae: Mustard, rape seed, radish, cabbage, cauliflower, knowl khol.

d) Cucurbitaceae: All gourds (Ridge, Ash, Bitter, Snake gourd), cucumber, pumpkin *etc.*

e) Malvaceae: Cotton, lady's finger.

f) Solanaceae: Potato, tomato, tobacco, chillies, brinjal.

g) Tiliaceae: Jute.

h) Asteraceae (Compositae): Sunflower, safflower, niger.

i) Chenopodiaceae: Sugarbeet.

j) Pedaliaceae: Sesame.

k) Euphorbiaceae: Castor, tapioca.

l) Convolvulaceae: Sweet potato.

m) Umbelliferae: Coriander, carrot.

n) Liliaceae: Onion, garlic.

o) Zingiberaceae: Ginger, turmeric.

4. Commercial classification

Based on the plants used for different purpose :

a) **Food crops:** Rice, wheat, greengram, soybean, groundnut, *etc.*

b) **Feed crops / Forage crops:** All fodders, oats, sorghum, maize, napier grass, stylo, lucerne *etc.*

c) **Industrial / Commercial crops**: Cotton, sugarcane, sugarbeet, tobacco, jute, *etc.*

d) **Food adjuncts:** Turmeric, garlic, cumin, *etc.*

5. Agronomic classification

This classification is based on use of crop plants and their products:

1. **Cereals:** The word cereal was derived from the Roman word '**Ceres**' which denotes a '**Goddess grains**'. They are cultivated grasses grown for their edible starchy grains (one seeded fruit-caryopsis). Larger grains used as staple food are cereals-rice, wheat, maize, barley, oats *etc.* Grain is a collective term for the fruit of cereals (caryopsis). The exception of cereal Buck Wheat (Polygone), it belongs to the family Polygonaceae is grouped as cereal.
2. **Millets:** Small grained cereals which form the staple food in hot drier climatic regions of the developing countries are called millets. Ex. Sorghum, Bajra, Ragi and Minor millets like Navane, Harka, Baragu, Same. They have single cover (due to the fusion of hesta to the pericarp).
3. **Oil seeds:** Crops that yield seeds that are rich in fatty acids, are used to extract vegetable oils. Ex. Mustard, Rapeseed, Sesame, Sunflower, Castor, Linseed, Groundnut, Soybeans.
4. **Legumes or Pulses:** Legume derived from Latin word '**Legre**' meaning '**To gather**' because the pods have to be gathered or picked by hand. These are belongs to the family Leguminaceae, form nitrogen fixing nodules on their roots. The pulses after splitting produce Dal which is a rich source of proteins. Ex. Greengram, Blackgram, Redgram, Bengalgram, Soybean, Cowpea, Peas, Lentil.
5. **Feed / Forage:** It refers to vegetative matter, fresh or preserved, utilized as feed for animals. It includes hay, silage, soilage, pasturage and fodder of cereals and legumes. Ex. Baja, Napier Grass, Guinea Grass, Fodder sorghum, Fodder maize, Lucerne *etc.*

6. **Fibre crops:** Plants grown for their fibre yield. There are different kinds of fibre. Any thread like material / tissue having sufficient toughness for use in textile and similar industries are fibre. They are i) seed fibre-Cotton, ii) stem / bats fibre-Jute, Linseed, Mesta, iii) leaf fibre-Agave, Pineapple.
7. **Sugar and starch crops:** Crops grown for production of sugar and starch. Ex: for Sugar -Sugarcane and Sugarbeet. For starch-Potato, Sweet Potato Tapioca.
8. **Spices and Condiments:** Crop plants or their products used to season, flavour, taste, and add colour to the fresh or preserved food. Ex. Ginger, Garlic, Fenugreek, Cumin, Turmeric, Chillies, Onion, Coriander.
9. **Drug crops / Medicinal plants:** Crops used for preparation of medicines. Ex. Tobacco Mint *etc.*
10. **Narcotics:** Plants / products used for stimulating, numbing, drowsing or relishing effects. Ex. Tobacco, Ganja, Opium.
11. **Beverages:** Products of crops used for preparation of mild, agreeable and stimulating drinking. Ex. tea, coffee, cocoa.
12. **Dye crops:** Used for synthesis of dyes. Ex. Indigo, Safflower *etc.*

6. Seasonal classification

Crops are grouped under the seasons in which they cultivate.

a) ***Kharif*** **or South West Monsoon season crops:** Crops grown during June-July to September-October which requires a warm wet weather during their major period of growth and shorter day length for flowering. Ex. Rice, Maize, Castor, Groundnut.

b) ***Rabi*** **Crops:** Crops grown during October-November to January-February which require cold dry weather for their major growth period and longer day length for flowering. Ex. Wheat, Mustard, Barley, Oats, Potato, Bengalgram.

c) ***Zaid* or summer crops:** Crops grown during February – March to May-June which requires warm dry weather for growth and longer day length for flowering. Ex. Blackgram, Greengram, Sesame, Cowpea *etc.*

7. According to ontogeny

It is a classification based on the life cycle of a plant.

a) **Annual Crops:** Crop plants that complete life cycle within a season or year. They produce seed and die within the season. Ex. Wheat, Rice, Maize, Mustard.

b) **Biennial Crops:** Plants that have life span of two consecutive seasons or years. First year / season these plants have purely vegetative growth. During the second year / season they produce flower stocks from the crown and after producing seeds the plants die Ex. Sugarbeet, Beet root, Cabbage, Carrot, *etc.*

c) **Perennial Crops:** They live for three or more years. They may be seed bearing or non-seed bearing. Ex. coconut, Arecanut, Napier grass. In general perennial crops occupy land for more than 30 months.

8. According to cultural requirement of crops

Certain group of plants are alike in cultural requirements due to their similar agro-botanical or morpho-agronomical characters.

i) **High land / dry land crops:** Crops that cannot tolerant water stagnation. Ex. Redgram, Groundnut, Maize, Sorghum, Cotton, Sesame, Napier *etc.*

ii) **Medium / mid land / garden land crops:** Crops that require sufficient soil moisture but cannot tolerate water stagnation. Ex. Potato, Sugarcane, Upland Rice, Ragi, Wheat, Blackgram, Bengal gram.

iii) **Low land / Wetland crops:** Crops that require abundant supply of water and can withstand prolonged water logged conditions. Ex. Rice, Dhaincha.

9. Based on textural suitability of soils

i) **Crops suitable to sandy to sandy-loam (light) soils:** Sorghum, bajra, greengram, sunflower, potato, onion, carrot *etc.*

ii) **Crops suitable to silty loam (medium) soils:** Jute, sugarcane, maize, cotton, mustard, tobacco, bengal gram, redgram, cowpea, *etc.*

iii) **Crop suitable to clay to clay loam (heavy) soils:** Rice, wheat, barley, linseed, lentil, paragrass, guinea grass *etc.*

10. Based on tolerance to problem soils

i) Tolerant to acidic soils: Wet rice, potato, mustard.

ii) Tolerant to saline soils: Chilli, cucurbits, wheat, sorghum, bajra, cluster beans, barley *etc.*

iii) Tolerant to alkali /sodic soils: Barley, cotton, bengalgram, berseem, sunflower, maize, *etc.*

iv) Tolerant to waterlogged soils: Wet rice, dhaincha, paragrass, napier grass, guinea grass.

v) Crops tolerant to soil erosion: Marvel grass, groundnut, blackgram, rice bean, horsegram.

11. According to cultivation of crops

i) **Arable Land -** Land suitable for cultivative of field or horticulture crops.

ii) **Non-arable land -** Land which is unsuitable for arable farming and usually has limitation of water, Soil slope, problematic soils, climate, *etc.* Rocky and mountain or forest land.

12. According to the depth of root system

i) Shallow rooted crops (< 30 cm): Rice, Potato, Onion.

ii) Moderately deep rooted (30-45 cm): Wheat, Groundnut, Castor, Tobacco.

iii) Deep rooted (45-60 cm): Maize, Cotton, Sorghum.

iv) Very deep rooted (> 60 cm): Sugarcane, Safflower, lucerne, redgram.

13. According to the tolerance to hazardous weather condition

i) Frost tolerant: Sugarbeet, beet root.

ii) Cold tolerant: Potato, cabbage, mustard.

iii) Drought tolerant: Bajra, jowar, barely, safflower, castor.

14. According to water supply

i) Irrigated crops: Rice, wheat, pulses, groundnut, berseem.

ii) Rainfed / upland crops: Jute, maize, ragi, upland rice, redgram.

iii) Rainfed but partially irrigated: Bengal gram, wheat, jowar, bajra, mustard, horsegram.

iv) Residual/conserved soil moisture: Rapeseed, bajra, barley, safflower, linseed.

v) Rainfed with supplemental irrigation: *Kharif* rice, sugarcane, blackgram.

vi) Rainfed with flooded water: Deep water rice, jute, sugarcane, dhaincha.

15. According to method of sowing / planting

i) Direct seeded crop: Upland Rice, Wheat, Ragi, Jowar, Bajra, Groundnut etc.

ii) Planted crops: Sugarcane, Potato, Sweet Potato, Napier, Guinea grass.

iii) Transplanted crop (after raising seedlings in the nursery): Rice, ragi, tobacco, brinjal.

16. Based on length of duration of crops

i) Very short duration crops: up to 75 days Pulses.

ii) Short duration crops: 75-100 days Sunflower, cauliflower, upland rice.

iii) Medium duration crops: 100-125 days Wheat, jowar, bajra, groundnut, jute.

iv) Long duration crops: 125-150 days Mustard, tobacco, cotton.

v) Very long duration crops: above 150 days Sugarcane, redgram, castor.

17. Based on method of harvesting

i) Reaping: Rice, wheat.

ii) Uprooting by pulling: Bengal gram, black gram, lentil, rapeseed.

iii) Uprooting by digging: Potato, sweet potato, groundnut, carrot *etc.*

iv) Picking: Cotton, vegetables, brinjal, bhendi, chillies.

v) Priming: Tobacco.

vi) Cutting: Berseem, napier, amaranthus.

vii) Grazing: Stylosanthes.

18. According to post harvest requirement

i) Curing: Tobacco, mustard

ii) Stripping: Jute, sunnhemp

iii) Shelling: Groundnut

iv) Ginning: Cotton

v) Seasoning: Turmeric, chillies

vi) Grading and sorting: Potato, rice, wheat, fibre crops *etc.*

19. Based on climatic condition

1. Tropical crop: Coconut, sugarcane
2. Sub-tropical crop: Rice, cotton
3. Temperate crop: Wheat, barley
4. Polar crop: All pines, pasture grasses.

20. According to important Uses

Though plants are useful in many ways only certain uses are given below:

1. **Catch crops / contingent crops:** Are those crops cultivated to catch the forth coming season. It replaces the main crop that has failed due to biotic or climatic or management hazards. Generally, they are of very short duration, quick growing, harvestable or usable at any time of their field duration and adaptable to the season, soil and management practices. Ex. Greengram, Blackgram, Cowpea, Onion, Coriander, Bajra.
2. **Restorative crops:** Are those crops which enrich or restore soil fertility or amelioration of the soils. They fix atmospheric nitrogen in root nodules, shed their leaves during ripening and thus restore soil conditions. Ex. Legumes.
3. **Exhaustive crops:** Are those crop plants which on growing leave the field exhausted because of a more aggressive nature. Ex. Gingelly, Maize, sunflower *etc.*
4. **Paira crop / residual crops:** Are those crop plants which are sown a few days or weeks before the harvest of the standing mature crops to utilize the residual moisture, without preparatory tillage. The standing crop and the later sown (paira) crop become simultaneous (forming a pair) for a short period. For example Black gram in Paddy
5. **Smother crops:** Are those crop plants which are able to smother or suppress the weed growth by providing suffocation (curtailing movement of air) and obscuration (of the incidental radiation) Ex. Cowpea.

6. **Cover crops:** Are those crop plants which are able to protect the soil surface from erosion (wind, water or both) through their ground covering foliage and or root mats. Ex. Groundnut, blackgram, marvel grass, sweet potato.
7. **Nurse crops:** A companion crop which nourishes the main crop by way of nitrogen fixation and or adding the organic matter into the soil. Ex. Cowpea intercropped with cereals, glyricidia in tea.
8. **Guard / barrier crops:** Are those crop plants which help to protect another crop from trespassing or restrict the speed of wind and thus prevent crop damage. Main crop in the centre surrounded by hardy or thorny crop. Ex. Mesta around sugarcane; sorghum around cotton; safflower around gram.
9. **Trap crops:** Are those crop plants grown to trap soil borne harmful parasitic weeds. Ex. Orobanche and striga are trapped by solanaceous and sorghum crops respectively. Nematodes are trapped by solanaceous crops (On uprooting crop plants, nemathodes are removed from the soil). Ex. Castor in cotton.
10. **Augmenting crops:** Are those sub crops sown to supplement the yield of the main crop. Ex. Mustard or cabbage with berseem to augument the forage yield of berseem.
11. **Green manure crops:** Grown & incorporated freshly in to the soil to increase soil fertility. It may be (a) Green leaf manuring – Calotrophis, (b) Green manuring *Insitu* – Sunhemp.
12. **Silage crop:** These are grown to preserve in pits in a succulent condition by a process of natural fermentation or acidification for feeding livestock during lean months Ex. Maize, sorghum, Cowpea.
13. **Soilage crop:** These are grown to harvest while they are still green and fed fresh to livestock Ex. Cowpea, Napier, Horsegram.

Chapter 7

Seeds and Sowing

Seed is the living link between plant and progeny and biologically seed is ripe fertilized ovule and a unit of reproduction of flowering plant. Agronomically, a seed material or propagule is the living organ of crop in rudimentary form used for propogation.

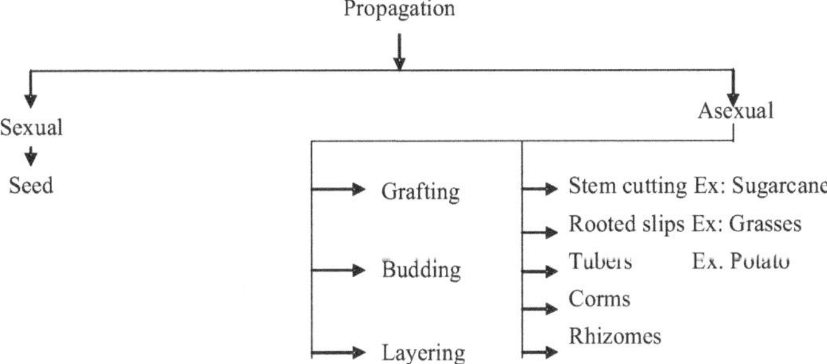

Plants perpetuate / multiply either sexually or asexually. Sexual methods are through seeds and asexually through vegetative parts.

Most of the crop plants produce viable seeds which are used for sowing. Those used for multiplication are called seeds and those which used for human consumption are called grains.

Seed consisting of intact embryo, stored food and seed coat which is viable and has got a capacity to germinate. Embryo has two parts:

a) **Plumule:** Which grows upward as shoot

b) **Radicle:** Grows downwards as roots

In India, seed multiplication program is three generation namely breeder, foundation and certified seeds as it flows from breeder to farmers.

Nucleus seed: Is the initial seed obtained from selected individual plant of a particular variety or parental line for the purpose of purifying and maintaining that variety or parental line by the organizing breeder and it is further multiplied under his own supervision or a qualified plant breeder. It is used to produce breeder seed and forms the basis of the total seed production chain.

Breeder seed: Is the progeny of nucleus seed of a variety and is produced by organizing breeder by a sponsored breeder.

Foundation seed: Is the progeny of breeder seed and is required to be produced from breeder seed or from foundation seed which can be clearly traced to breeder seed.

Certified seed: Is the progeny of foundation seed and must meet the standards of seed certification prescribed in the minimum seed certification standards 1988.

Characteristics of a good seed

Good quality seed must be :

- True to type
- Healthy, pure and free from inert materials and weed seeds
- Viable and germination capacity is up to the standard
- Uniform in texture, structure and appearance
- Shouldn't affected by any seed born disease / pest.

Seed germination: Germination is the development of seedling from the seed embryo which is able to produce a normal plant under favourable condition.

Types of germination: Based on cotyledon character, the germination is grouped as

1. **Epigeal germination:** In dicotyledons / pulses, the cotyledons are brought out of the soil by curved elongation of hypocotyls. This type of germination where cotyledons emerge out of the soil is called epigeal germination.

2. **Hypogeal germination:** In monocotyledons / cereals, the cotyledon remain in the soil, the plumule grows or is pushed upward by the elongation of epicotyl. This type of germination is called hypogeal germination.

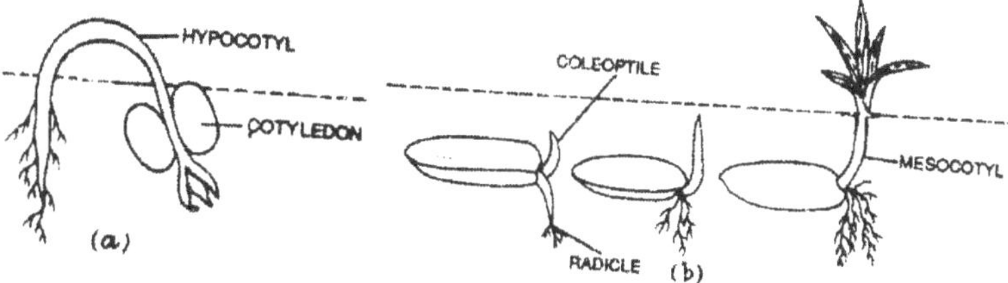

Factors influencing seed germination

Generally, conditions favouring seedling growth also favour germination. These factors are grouped into two :

1. Seed or inherent factors
2. External or environmental factors.

1. **Seed or inherent factors:** Seed viability and dormancy are two inherent factors. Viability of seeds represents the capacity of the seeds to germinate. Normally, a viable seed germinates in about a week of sowing.

Seed viability is defined as the degree to which a seed is metabolically active and capable of germinating under favourable condition which is governed by :

i) Storage condition of the seeds
ii) Age of the seeds
iii) Genetic constitutes of seeds
iv) Environmental conditions prevailing at maturity

Seed dormancy is the state of rest period of seed in which it doesn't germinate even under favourable conditions:

i) Hard seed coat Ex: Amaranthus, Cotton
ii) Seed coat being impermeable to oxygen Ex : Xanthium

iii) Rudimentary embryo of seeds Ex. Orchid seeds

iv) Dormant embryos Ex. Apple, Peach, Pinus

v) Synthesis and accumulation of germination inhibitors in the seeds.

Kinds of seed dormancy

a) **Primary dormancy:** The seeds which are incapable to germinate immediately after ripening / maturity even if we provide all the factors favourably. Ex. In Potato, individual eye bud is dormant for 7-11 weeks.

b) **Secondary dormancy:** The seeds are capable of germinating after ripening. But, due to un favourable conditions at storage for few days make them incapable to germinate.

Types of dormancy

i) **Physical dormancy:** It may be due to the presence of impermeable or mechanically resistant seed coat which affects water uptake and gaseous exchange.

ii) **Physiological dormancy:** It may be due to

   a) Presence of immature embryo

   b) Need for specific light after ripening and temperature

   c) Presence of germination inhibitors

   Seed dormancy can be overcome through following methods

Methods of dormancy breaking

A) **Physical treatment or Stratification:** Here, hard seed coat is broken by heat or pressure:

   a) Heat treatment at 40-45°C for different durations

   b) Low temperature treatment at 2-8°C for 12-24 hours

   c) Leaching of germination inhibitors through soaking seeds in running tap water

   d) Alternate heating and cooling

e) Alternate wetting and drying

f) Exposing water soaked seeds to red light for 1-2 hrs at 15-25°C.

g) Dehusking or removal of seed coat by beating

B) **Chemical treatment or Scarification:** Hard seed coat is broken by chemical means:

a) By acid treatment: Solutions of HNO_3, HCl, H_2SO_4

b) By gasses: Increasing concentration of oxygen

c) Hot water: 75-100°C

d) Non-harmonal substances: Thiourea, Ascorbic acid, KNO_3 *etc.*

e) Harmonal: GA, Kinetin *etc.*

Seed treatment: Defined as treating the seeds with different substances *viz.*, water, fungicide, growth regulator, nutrient substances or any other physical treatments aiming at

a) Breaking of dormancy Ex. Physical or Chemical treatment.

b) Improving germination & emergence Ex: Soaking in water, hormones.

c) Protection from insect pests and diseases including storage pests.

d) Improving seedling vigour and promoting initial crop growth.

e) Supplying nutrient & inoculation of bio-fertilizers.

f) Inducing drought resistance.

Establishment of field crops may be grouped into

1. Direct sowing 2. Planting 3. Transplanting

1. **Direct sowing:** Sowing is the placement of seed in the seedbed at appropriate depth and distance where the soil environment is

ideal for optimum germination and crop stand establishment. Efficiency of sowing depends on time, depth and method of sowing.

A) **Time of sowing:** Optimum time of sowing for each crop has been well established for different agro-climatic zones. The basic principle behind optimum sowing time is that the different phenophases of crop should coincide with optimum weather to yield remunerative crop. Different cultivars of the same crop may respond differently to the sowing time.

Delayed sowing invariably reduces yields attributing to the exposure of crop to different stresses *viz*., temperature, light, rainfall, pest and diseases *etc*. However, optimum sowing time has not practical under rainfed/dryland situation attributing to the unpredictable rainfall.

B) **Depth of Sowing:** Depth of sowing is another important aspect for establishment of good crop stand. Uneven depth of sowing results in uneven crop stand. Too shallow or too deep sowing leads to lower plant population owing to poor germination of seeds. Under such circumstances, crop appears with uneven spread with many gaps and weeds become problematic. The optimum depth of sowing depends on size, seed food reserve, coleoptile length and soil moisture.

Crops with bigger seeds like ground nut, castor, sunflower *etc.* can be sown upto depth 6-10 cm. Small sized seeds like ragi, tobacco have to be placed as shallow as possible. Shallow sown small sized seeds should water frequently. Otherwise, the soil and seed dries quickly and seed looses viability / fails to germinate. Deep sown seeds must have sufficient seed reserves to put forth long coleoptile for emergence.

Sowing depth is also determined based on mode of germination. In hypogeal germination, the cotyledons remain in the soil. Such seeds if sown deep, either delays or retards germination. While in epigeal mode of germination the cotyledons emerge out of the soil. Such seeds if sown deep, will fail to germinate.

The thumb rule is to sow seeds to a depth approximately 3-4 times their diameter. Optimum depth of sowing for most of the crops is 3-5 cm. Shallow placing (2-3 cm) is recommended for small sized seeds like ragi, bajra *etc*. Very small seeds like tobacco are placed at 1 cm depth.

C) **Methods of sowing / planting:** There are different methods of direct sowing as given below:

a) **Broadcasting:** Seeds are spread uniformly over well prepared land and covered by planking. It is practiced in dryland area. It is the most primitive method, easier and quicker. It has several disadvantages like:

- Difficult to maintain uniform plant population because of uneven placement
- Needs skilled labour to distribute the seeds uniformly
- High seed requirement
- Difficult in carrying intercultural operation
- Weeds become serious issues

b) **Dibbling:** Here, seeds are placed / dibbled manually in the holes / lines marked by running markers. Though this method is laborious and time consuming, it gives rapid and uniform germination and requires less seed rate than broadcasting.

c) **Plough sole placement:** Here seeds are placed in the furrow opened during ploughing and subsequently covered in the next turn. Bold seeds like groundnut are sown in this method. Optimum moisture is essential here to maintain uniform depth.

d) **Drilling:** To overcome the problem associated with broadcasting, seeds can be dropped uniformly in the furrows using seed drill. Animal or power operated seed / seed cum fertilizer drills can be used. This is one of the best methods that provides uniform plant establishment and weeds can be managed through intercultivation compared to broad casting.

2. **Planting:** Seed materials in vegetative propagated plants *viz.,* sugarcane (Setts), potato (Tubers), turmeric (Corms) *etc.,* are placed / planted at required distance.

3. **Transplanting:** Transplanting is the planting of seedlings in the main field raised in nursery unit. This method is adopted in crops where,

   a) Plants can able to withstand transplanting shock (Condition in which the plant is devoid of nutrient / water due to the break of soil-plant root interaction).

   b) When more number of crops to be taken in a year.

   c) Limited availability of time especially in drylands.

   d) Crops which demand extra care (moisture / pest & disease etc.) during seedling stage due to their tenderness which might difficult on large area of main field.

   e) The seedling growth is very slow in the early stage.

Under the above circumstances, seeds are sown in a small area called **Nursery** and seedlings are raised with all necessary cares. When they attain certain stage, they are pulled out and transplanted in the main field. It has certain advantages like water saving, good establishment and increase the cropping intensity.

For achieving good results from transplanting, the seedlings are to be transplanted at optimum time and at proper depth. The thumb rule for optimum age of seedlings is one week for every month of total duration. The depth of planting must be shallow in tillering crops to promote good tillering.

Chapter 8

Soil and Its Components, Fertility Productivity and their Management

Soil is defined as a thin layer of earth's crust which serves as a natural basic medium for the plant growth. Soil is a dynamic natural body composed of minerals, organic materials and living forms on which plants grow. Soil are formed by the weathering of rocks and minerals in presence of soil forming factors like climate, organisms, relief or topography, parent material and time (cl,o,r,p,t).The weathering of rocks and minerals are associated with physical and chemical phenomenon. The principal agents of physical weathering are temperature, water, wind, plant and animals. Chemical weathering is affected through the process of hydrolysis, hydration, carbonation, oxidation and reduction.

Each soil is characterized by a given sequence of horizons. Combining of this sequence is known as soil profile. **Soil profile** is a vertical section of the soil through its horizons. The layers or horizons in the soil profile vary in thickness and have different characteristics like colour, texture, structure etc. The horizons are designated by O, A, E, B, C and R.

Horizon: The layers seen in the vertical section of the soil profile are called horizons.

Soil is composed of 2 major components

1. **Solid phase:** 45 per cent inorganic and 5 per cent organic.
2. **Pore Space:** 25 per cent air and 25 per cent water.

Soils widely vary in their characteristics and properties like physical, chemical and biological properties. Understanding the properties of soil is important for their optimum use and best management practices.

Physical properties: Physical properties of soil are texture, structure, density, colour, porosity, plasticity, consistency *etc.*

1. Texture: It refers to the relative proportions of primary soil separates that is sand, silt and clay that are present in soil. These soil separates are grouped into 7 groups according to USDA (United States Development Agency)

a) Very course sand: > 1.0 mm
b) Course sand: 0.5 to 1.0 mm
c) Sand: 0.25 to 0.50 mm
d) Fine sand: 0.10 to 0.25 mm
e) Very fine sand: 0.05 to 0.10 mm
f) Silt: 0.02 to 0.005 mm
g) Clay: < 0.002 mm

As per the international union of soil science soil separates are grouped into 4 groups which are known as international method & are widely accepted.

a) Coarse sand: 0.2 to 2.0 mm
b) Fine sand: 0.02 to 0.2 mm
c) Silt sand: 0.002 to 0.02 mm
d) Clay: < 0.002 mm

A large amount of sand in a soil will make a coarse texture and gritty. Such soils are called Sandy or Sandy loam and are easy to work. When silt is present in large amount, soils feel like flour and are called silty loam.

Clay in a soil, make it sticky when wet and hard when dry resulting into clay or clay loam soils. A fine textured soil is made of largely of silt and clay. Its plasticity and stickiness indicates difficult to work or heavy.

The term light and heavy refers to ease to work and not by weight. The weight of sand is greater than clay but, the clay is heavy soil and sand is light soil.

Various textural classes that are found in soil are given under the table

Group	Texture	Sand (%)	Silt (%)	Clay (%)
Coarse Textured	Loamy sand	85	10	5
	Sandy loam	65	25	10
Medium textured	Loamy	45	40	15
	Silt loam	20	60	20
Fine textured	Silty clay loam	15	55	30
	Clay loam	28	37	35
	Clay	25	30	45

Textural management of soil

1. **Course textured soil / Sandy soils**
   a) These are loose, friable, posses good aeration & Drainage
   b) Percolation losses are more
   c) Low in nutrient & Water holding capacity.

Management: Application of organic matter to improve aggregation.

2. **Fine textured soil / silty / clayey soil**
   a) They are often called as hard or heavy soil.
   b) These soil have high plasticity & cohesion.
   c) They are high in water & nutrient holding capacity.
   d) Has poor drainage & infiltration and is subjected to water logging. They tend to big clods if ploughed when dried.

Management

a) Timely tillage operation.

b) Addition of organic manure.

c) Cultivation of deep rooted legumes.

Loamy soils seem to be ideal for crop production. A ideal loam may be defined as a mixture of sand, silt and clay particles which exhibits light and heavy properties in equal proportion. Loams posses the desirable characters of both sand and clay such as poracity, water and nutrient holding capacity, cohesion and plasticity *etc*.

2. Structure: The arrangement of soil particles is called soil structure. It influences aeration, permeability *etc.* There are 4 types of primary structures.

a) **Platy:** Horizontal arrangement of particles around a plane is called platy structure.

b) **Prism like:** Particles are arranged around the vertical axis and it may be of 2 types

   i) **Prismatic:** It has prism like outer surface

   ii) **Columnar:** It is like column.

c) **Block like:** Flat or rounded surface form block like structure. It may be

   i) **Angular blocky:** They have definite angle

   ii) **Sub-angular blocky:** Irregular outer surface

d) **Spheroidal:** are bound by curved or spherical surface and are most favourable for tillage and root penetration. It may be:

   i) **Granular:** like grains

   ii) **Crumb like:** Porous

Factors affecting soil structure

1. **Soil management:** Ploughing and other tillage operations may break the soil particles and alter soil structure. Use of green

manures and legumes & application of irrigation water also influence soil structure.

2. **Adsorbed cations:** Na^+ and K^+ are dispersing elements while Ca^{2+} and Mg^{2+} flocculate the soil particles.

3. **Application of organic matter:** Improves soil aggregation.

3. Soil colour: The colour of the soil is mostly due to the presence of Fe and Mn compounds and organic matter in the soil. Organic matter makes the soil dark brown or grey. Oxides of Fe and Mn impart red, brownish red or yellow colour depending on the oxidation status and extent of hydration. Soil colour has a direct influence on soil temperature. Darker soil absorbs more heat and warm up quickly. Darker soils are fertile and give better yield. Soil colour is measured by Munsell Chart.

4. Density: The weight of a substance per unit volume is called density. The soil density is expressed in 2 ways that is bulk density and particle density.

S. No.	Particle density (PD)	Bulk density (BD)
1.	It is defined as weight of unit volume of soil solids excluding pore space $\text{Particle density} = \frac{\text{Weight of soil}}{\text{Volume of solid}}$	Weight of unit volume of dry soil which includes both solid and pore space $\text{Bulk density} = \frac{\text{Weight of soil}}{\text{Volume of soil (Solid + Pores)}}$
2.	It is not affected by pore space	Largely affected by pore space
3.	——— – 2.75 Mg m^{-3} or g cc^{-1}	The value normally vary between 1.40 – 1.80 Mg m^{-3} or g cc^{-1}
4.	It is calculated by compressing all solids to bottom If the soil of 1 cc weighs **1.33** g 0.5 cc Pore space =1.33/0.5 **0.5 cc** Solid = **2.66 g cc**$^{-}$ 1	It is calculated by taking the total volume 1 0 cc $BD = \frac{1.33}{1.0}$ = **1.33 g cc**$^{-1}$

5. Pore space: The pore space of a soil is that portion that occupied by air & water. If the soils particles are tend to be close as in sand, the total porosity is low. If they are arranged in porous aggregates as in medium & fine textured soil, pore space per unit volume is high. The pore space is calculated by:

$$\frac{BD = \text{Weight of soil / Volume of soil}}{PD = \text{Weight of soil / Volume of solid}}$$

$$\frac{\text{Volume of solid}}{\text{Volume of soil}} = \text{Solid space}$$

$$\% \text{ Solid space} = \frac{BD}{PD} \times 100$$

% Pore space + % Solid space = 100

% Pore space = 100 - % Solid space

% Pore space = 100 - [BD/PD X 100]

% Pore space = 100 (1 – BD/PD)

Problem 1. A sandy soil of 10 cc weighs 15 g and reduced to 6 cc after compression. Calculate BD, PD and % Pore space?

BD = 1.5 g/cc PD = 2.5 g/cc % PS = 40 per cent

Problem 2. A silty loam soil having BD =1.3 g/cc & PD =2.6 g/cc. Calculate the % pore space

% PS = 50 per cent

Sand particles have 35-50 per cent pore space and clay particles have 40-60 per cent pore space. Based on the size, the pores are grouped as

a) Macro pores: >200 μm – Space for air and

b) Micro pores: < 200 μm – Space for water

Macro pores allow ready movement of air & water while micro pores impede the movement.

Sandy soils are largely composed of macro pores while, clay soils have more micro pores. Hence water holding capacity of clay soils is higher.

Soil type	% Pore space
Sandy soil	20-30
Loamy soil	30-50
Clay soil	50-60

6. Plasticity & Cohesion: Plasticity is the soil property, which enables a moist soil to change shape upon applied force and retain this shape even when the force is withdrawn. Sandy soils are non-plastic and clayey soil is plastic. Adhesion is the binding property of different particles (soil with others – Water, nutrient). Cohesion is the binding property of similar molecules.

Plasticity, adhesion & cohesion together reflect soil consistency and workability. Black soils are highly plastic and adhesive making tillage operation difficult.

7. Chemical Properties: Soil reaction and electrical conductivity (EC) are two important chemical properties affect plant growth.

I. **Soil reaction / Soil pH:** The soil pH indicates its acidity or alkalinity and influence nutrient availability, soil physical condition and inturn crop growth. To soil pH influences the following.

   a) **Nutrient availability:** Soil pH influences rate of nutrient released through decomposition and solubility of minerals. The source of nitrogen and sulphur is organic matter which is released during its decomposition. The rate of decomposition is fastest between 6-8 pH. Phosphorous is best available between 6.5-7.5. Less than 6.5 precipitate P by Fe & Al and more than 8.5 through Ca & Mg. Fe, Mn, Cu, Zn, B precipitates at higher pH. Optimum pH range for availability of nutrient is :

S.No.	Nutrient	Opt. pH	Sl. No	Nutrient	Opt. pH
1.	N	6.0-8.0	6.	Fe	< 6.0
2.	P	6.5-7.5	7.	Mn	5.0-6.5
3.	K	6.0-7.5	8.	B, Cu, Zn	5.0-7.0
4.	Ca, Mg	7.0-8.5	9.	Mo	> 7.0
5.	S	6.0-7.5			

   b) **Soil physical condition:** Soil physical condition like flocculation, colour *etc.* depends on soil pH. Higher soil pH

(>8.5), increases Sodium content deflocculates soil colloids and destruct soil structure & movement of water & air. Under acidic condition, Fe dissolved in excessive amount impart red colour to the soil.

II. **Electrical Conductivity (EC):** The concentration of soluble salts in soil can be termed as electrical conductivity (EC) and expressed as dSm^{-1} (dec siemens per meter). Salts like sulphates. Chlorides, carlo onates & bicarbonates of Ca, Mg, Na, K and Nitrates of Ca, Mg, Na, K.

S.No.	EC dSm^{-1}	Salivity class
1.	< 0.8	Low
2.	0.8 to 1.6	Medium
3.	1.6 to 2.5	High
4.	> 2.5	Very High

Soil fertility and Productivity

Soil fertility is the inherent capacity of the soil to supply nutrients in adequate amounts and in suitable proportion. Soil productivity is the capacity of the soil to produce crops with specific management and expressed in terms of crop yield.

All productive soils are fertile, but all fertile soils need not to be productive. It may be due to some problems like water logging, saline or alkaline condition, weeds, pest and diseases etc. under these conditions, the crop growth is restricted even though the soil has sufficient amount of nutrient.

Factors affecting soil fertility: The factors governing soil fertility may be:

i) **Natural or soil farming factors:** Include climate, topography, relief, parent material and time or age (cl,t,r,p,t).

ii) **Artificial & soil management factors:** Addition of organic and inorganic fertilizers, crop rotation and other soil management practices.

Soil fertility evaluation

Several techniques are used to assess soil fertility status. Important ones are:

a) **Visual nutrient deficiency symptoms:** Looking to the characteristic symptoms on plants.

b) **Plant analysis or tissue test:** Analyze the nutrient content and match it with critical limit. It is also called crop logging, developed by Clements in Sugarcane.

c) **Biological test:** Culturing micro-organisms or growing plants on sample soil.

d) **Chemical method or soil test:** Actual soil analysis for different nutrient.

Soil fertility	Soil productivity
It is an index of nutrient availability	It is broader term to indicate crop yield.
It is one of the factors of crop production.	It is the interaction of all the factors that determine crop yield.
Howevercrop production also depends on water, slope, depth of water tale etc.	
It can be analyzed in the laboratory	It is assessed under field condition.
It is expressed as nutrients in kg ha^{-1}	It is expressed in terms of crop yield.

Chapter 9

Tillage and Tilth

Cultivation involves management of physical environment to produce a favourable habitat for successful crop production. Primitive man used tools to disturb the soil for placing the seeds. The word tillage is derived from the Anglo-Saxon words ***tilian*** and ***teolian*** means to **plough** and prepare soil for seed sowing, to cultivate and to raise crops. Jethrotull who is considered as father of tillage suggested that ploughing is necessary so as to make the soil into fine particles.

Tillage is the mechanical manipulation of the soil either with tools and implements for obtaining conditions ideal for seed germination, seedling establishment and growth of crops.

Tillage includes all operations and practices that are used for the purpose of modifying soil physical characters. Tillage is the most difficult and time consuming operation in crop production. About 30 per cent of the total cost of cultivation goes for tillage operations.

Objectives of tillage

1. To prepare good seed bed which helps in germination and establishment of seedlings.
2. To provide adequate seed-soil contact to permit water flow to seed and seedlings.
3. To incorporate plant residues, manures and fertilizers uniformly in the soil.
4. To make the soil capable for absorbing more rain water and minimizing soil erosion.

5. To establish specific surface configurations for sowing, irrigation, drainage *etc.*
6. To incorporate and mix the applied fertilizers into the soil.
7. To expose weeds, soil borne pathogens to heat and killing them. To aerate the soil.
8. To remove the hard pan and to increase the soil depth.

Disadvantages of tillage

1. Unless the soil, moisture content is optimum, desired seedbed may not be possible. If the heavy soil are tilled at high soil moisture content, the soil gets puddle and may not be suitable for immediate sowing. If they are ploughed when too dry, big clods are turned up which are difficult to work into a good seed bed.
2. Repeated tillage operations hasten oxidation of organic matter from soil.
3. Tillage with heavy equipment breaks down the stable soil aggregates and form hard pan leading to low infiltration and induces run off, soil erosion and other negative effects.
4. Frequent tillage pulverizes the soil into dust, break down the soil aggregates and enhances wind and water erosion.
5. Repeated use of heavy machinery for tillage operations leads to soil compaction and development of dense root zone.

Tilth: Tilth is the term used to express the physical condition of soil resulting from tillage or it is optimum condition obtained out of tillage or it is the product or result of tillage. The tilth may be coarse tilth, fine tilth or moderate tilth based on the requirement of crops being grown and the soil where we are cultivating.

Influence of tillage on soil physical properties: Tillage has considerable influence on soil physical properties like pore space, structure, bulk density, soil water, soil colour and temperature. Hence, tillage influence germination, seedling emergence and crop stand establishment.

i) **Pore space:** Soil is made up of particles of different sizes. Air and water filled spaces between these particles constitute pore space. When a field is ploughed, the soil particles are loosely staked in a random manner and pore space increases.

Under good tilth condition, capillary and non-capillary pores are in equal proportion and facilitate free moment of air and water and increases infiltration.

ii) **Soil structure:** Soils with crumby or granular structure are good for crop production. When soil is subjected to tillage at optimum moisture leads to crumb like structure which reduces the erosion greatly. Soil structure will destroy when tillage is carried at inappropriate soil moisture.

iii) **Bulk density:** When the soil is loosened with tillage, the soil volume increases without any effect on weight. Therefore, the BD of tilled soil is less than untilled soil.

iv) **Soil colour:** Tillage increases the oxidation of iron and manganese, decomposition of organic matter resulting in fading of colour.

v) **Soil water:** The available soil water depends on soil porosity, soil depth and random roughness. Roughness is the measure of micro-elevations and depressions caused by ridges and furrows. All these characters are influenced by tillage. It also influences infiltration and water holding capacity.

vi) **Soil temperature:** Tillage loosens the soil surface resulting in decrease of thermal conductivity and increase heat exchange.

Types of Tillage: Tillage operations may be grouped into on- season and off-season tillage.

a) On-season tillage: Tillage operations for a crop from the start of crop season to the crop harvest are known as on season tillage operations. It includes primary tillage, secondary tillage and intertillage (inter cultivation).

1. Primary tillage / Ploughing

The operation that is done after the harvest of crop to bring the land under cultivation is known as primary tillage. Ploughing is the opening of the compacted soil with the help of different ploughs like country plough, mould board plough, bose plough, tractor and power tillers are used for primary tillage. Ploughing is done to cut open and invert the soil, uproot the weeds and stubbles.

The optimum time for ploughing is decided on soil moisture status. If the soil is ploughed when dry, (a) there will be formation of big clods (b) demands enormous energy and (c) it is difficult to operate. Too wet condition also unfavourable for ploughing as (a) the soil sticks to plough (b) there will be formation of hard pan in sub soil layer (c) the clods formed will be very hard after drying. Optimum range of soil moisture for effective ploughing is 25-50 per cent depletion of available soil moisture. Light soils can be ploughed in a wide range of soil moisture condition while, heavy soil have narrow range.

Depth of ploughing mainly depends on effective root zone. Tap rooted crops have deep roots and require deep ploughing while shallow fibrous roots demand shallow to medium ploughing.

The number of ploughings necessary to obtain good tilth depends on soil type, weed problem and crop residue on the soil. In heavy soils, 3-5 ploughings are necessary while, 1-3 ploughings are sufficient in light soils. When weeds and plant residues are higher, repeated (more) ploughings are necessary.

Selection of ploughs depends on the purpose, soil condition and nature of weed problems.

Types of primary tillage

i) **Deep Ploughing**: One centimeter of surface soil over one hectare of land weighs about 150 mt. Therefore, deep ploughing demands enormous energy. Cenral Research Institute for Dryland Agriculture (CRIDA), Hyderabad classified ploughing as :

1. Shallow ploughing – 5 to 6 cm depth.
2. Medium ploughing – 15 to 20 cm depth.
3. Deep ploughing – 25 to 30 cm depth.

Summer deep ploughing expose weeds and other soil borne pathogens to heat and turns out into large sized clods. The clods crumble due to alternate heating and cooling and due to occasional summer rains which will disintegrate clods gradually. Deep tillage improves soil moisture content due to enhanced wetting zone.

A deep tillage of 25-30 cm is necessary for deep rooted crops like pigeon pea, while moderate tillage of 15-20 cm is required for maize and wheat.

ii) **Sub soiling:** Sub soiling is the breaking of hard pans without inversion and with fewer disturbances to surface soil. Hard pans in the soils may be formed because of silt load, Fe & Al blocking, clay particles or man made will restrict the root growth of crops. Man made pans are induced because of repeated tillage at same depth or ploughing under excess soil moisture condition.

iii) **Year round tillage:** Tillage operations carried throughout the year are known as year round tillage. In dry land areas, field preparation is initiated with summer shower and repeated till sowing and even after harvest of crop, the field is repeatedly ploughed to avoid weeds and soil erosion.

2. Secondary tillage

The tillage operations that are performed on the soil after primary tillage to bring a good soil tilth are known as secondary tillage. Secondary tillage consists lighter or finer operations which are performed to clean the soil, break the clods and incorporate the manures and fertilizers. Harrows and planking is done to serve those purposes. Planking is done to crush the hard clods, level the soil surface and to compact the soil lightly.

3. Layout of seed bed

Certain crops like wheat, soybean, castor can be sown directly after secondary tillage, certain crops like maize, sugar cane, vegetables need ridges and furrows, certain crops can be sown with seed drills while, seeds of certain crops needs covering with harrows / plank / roller for necessary germination. It is essential to prepare ideal seed bed suiting to the particular crop for proper germination and establishment.

An ideal seed bed should provide the essential conditions for good germination and favourable environment for emerging seedlings. Such conditions include aeration, moisture and allow easy penetration of roots. It should also free for weeds and soil borne pathogens.

A firm seed bed may be required to ensure close contact between seed and soil whereas, a loose seed bed may be desirable for good aeration. Smaller the seed greater will be its sensitivity Ex. Tobacco to seed bed condition which demand fine seed bed. Crops with bold seeds are not demanding fine seed bed Castor.

4. Tertiary tillage /leveling

Leveling board, buck scrapers *etc.*, are used for leveling.

b) **After tillage (inter tillage / inter cultivation):** The tillage operations that are carried out in standing crop are after the sowing or planting and prior to the harvest of the crop plants are called inter tillage. It includes hoeing, weeding, earthling up, drilling or side dressing of fertilizers, aeration and to conserve moisture *etc.*

Implements used for different operations

i) Primary tillage: Wooden plough, sub soil plough, chisel plough, ridge plough, MB plough *etc.*

ii) Secondary tillage: Cultivators, harrows, plank, roller.

iii) Layout of seed bed: Country plough, seed drill.

iv) Inter-cultivation: Wooden plough, blade harrows, ridger, spade, hand hoes *etc.*

c) **Off season tillage:** Tillage operations done for conditioning the soil suitability for the forthcoming main season crop are called off-season tillage. This includes post harvest tillage, summer tillage, winter tillage and fallow tillage.

Dry tillage: Dry tillage is practiced for crops that are sown in dry land condition having sufficient moisture for germination of seeds. This is suitable for crops like broadcasting paddy, jute, wheat, oilseed crops, pulses, potato and vegetables. Dry tillage is done in a soil having sufficient moisture (21-23 %). The soil becomes more porous and soft due to dry tillage. Besides this, the water holding capacity (WHC) of the soil and aeration are increased.

Wet or puddling: The tillage operation that is done in a land with standing water is called **wet or puddling tillage**. Puddling operation consists of ploughing repeatedly in standing water until the soil becomes soft and muddy. Puddling creates an impervious layer below the surface to reduce deep percolation losses of water and to provide soft seed bed for planting rice, hastens transplanting operation and smoothly as well as establishment of seedlings.

Special purpose tillage: Tillage operations are intended to serve special purposes are said to be special purpose tillage and they are sub soiling, leveling, blind tillage, clean tillage and zero tillage.

1. **Zero tillage:** In this, new crop is planted in the residues of the previous crop without any prior soil tillage or seed bed preparation and it is possible when the weeds are controlled by the use of herbicides and also called as chemical tillage.

Zero tillage is practiced where

a) Soils subjected to erosion by wind and water.

b) Timing for tillage operation is too short or difficult.

c) Requirement of energy and labour is too high.

The organic matter content of soil increases under zero tillage due to slow mineralization. Zero tillage is practiced in Rice-wheat system in India.

In zero tillage, herbicide plays a dominant role in weed management. Before sowing, the vegetations are destroyed with broad spectrum, non-selective herbicides with relatively short residual effect (paraquat, glyphosate *etc.*), selective and moderately persistent herbicides (Atrazine, 2,4-D *etc.*) are required during subsequent stages.

2. **Clean tillage:** It refers to working of soil of the entire field in such a way no living plant is left undisturbed. It is practiced to control weeds, soil borne pathogens and pests.
3. **Blind tillage:** It refers to tillage done after seeding or planting the crop (in a sterile soil) either at the pre-emergence stage of the crop plants or while they may are in early stage of growth so that the crop plants (cereals, tuber crops *etc.*), don't get damaged, but extra plants and broad leaved weeds are uprooted.
4. **Minimum tillage:** It refers reducing tillage operations to the minimum necessary for ensuring a good seed bed, rapid germination, a satisfactory stand and favourable growing conditions. Tillage can be reduced by two ways.
   a) Omitting operations which don't give much benefit when compared to cost.
   b) Combining operations like seeding and fertilizer application.

Advantages of zero tillage

1. Improves soil conditions through reduced movement of heavy tillage implements, vehicles *i.e.*, less soil compaction.
2. Saves time and energy 50-58 per cent of reduction in fuel consumption.

Disadvantages of zero tillage

1. Poor germination of seeds and affect plant stand establishment.

2. Needs higher nitrogen attributing to slower decomposition of organic matter.
3. Affect nodulation in legumes.
4. Weeds are major problem & continuous use of herbicide cause pollution hazards.
5. **Conservation tillage:** The major objective is to conserve soil and soil moisture. It is systems of tillage in which organic residues (min 30%) are not inverted into soil that they remain on surface as protective cover against erosion and evaporation losses of soil moisture. If the stubbles form the protective cover on the surface it is usually **Stubble mulch tillage.**

Factors influencing tillage

Several factors influence the tillage are as follows:

a) **Crop:** It dictates the type and extent of tillage. Smaller seeds like finger millet, tobacco, *etc.*, require a fine seed bed which can provide intimate soil-seed contact as against relatively coarse seed be for big size seeds such as sorghum, maize, pulses, *etc.*, Delicate seeds such as tomato, chillies, cabbage and cauliflower require finer tilth with shallow cover over seeds. Root or tuber crops require deep tillage and rice needs puddling.

b) **Soil type:** It decides the range of soil moisture and the time (period) during which the soil can be tilled. Light soils require early and rapid land preparation due to free drainage and low retentive capacity as against heavy soils, which require considerable time to start tillage. The energy requirement of light soils is less than that of heavy soils. Loamy soils can be brought to good tilth with little expenditure.

c) **Climate:** It influences soil moisture content, draught required for tilling and type of cultivation. Low rainfall and poor water retentive capacity of shallow soil do not permit deep ploughing at the start of the season. Contour tillage is necessary for crops on sloppy lands. Heavy soils developing cracks during summer (self tilled) need only harrowing. Light soils of arid regions need coarse tilth to minimize wind erosion.

d) **Type of farming:** It influences the intensity of land preparation. In dry lands, where only one crop is taken per year, deep ploughing is necessary to eradicate perennial weeds and to conserve soil moisture. Land is covered with crops all through the year under intensive irrigated farming. Repeated shallow tilling is adequate under such intensive cropping.

e) **Cropping system:** Cropping system involving different crops need different types of tillage. In general, crop following rice needs repeated preparatory tillage for obtaining an ideal seed bed, tuber crops like potato require minimum tillage and pulses need lesser tillage than that of sorghum, maize and sugarcane.

Chapter 10

Crop Density and Geometry

Crop density is nothing but number of plants or seedling per unit area. Crop density per unit area depends on the soil, climate and type of crop being planted. For example under extreme conditions, in poor soils and in semiarid regions with no irrigation planting at low density is best. Otherwise crop plants will be thin and weak. In such conditions not only produces low yield but also is the ideal condition for pest and diseases. Crop unit area depends on :

1. Soil fertility: In poor soils, plant density should be lower than in fertile soils.
2. Availability of water: In areas where water is a limiting factor, planting should be done at a lower density.
3. Tillering capacity of the crop: Small grains and other cereals are planted at lower densities due to tillering capacity and elasticity of the crops.

Plant factors influencing crop density

Plant size: Plant size (leaf area) determines the number of plants required to develop optimum leaf are index for higher yield. At low crop density, crop canopy is open and LAI is low in such situations leaf size should be more. Generally bigger the plant size lesser will the crop density for optimum yield. Crops such as cotton, sunflower, safflower, maize and pigeonpea, *etc.*, require low crop density compared sorghum, rice, ragi, *etc.*

Tillering: All the cereals show the ability to compensate for low crop density by producing more tillers. Consequently yield may not be affected by seed rates over a wide range. Similarly, crops such as Cotton, Pigeonpea, Safflower and several other indeterminate crops can compensate for low crop density by development of more number of branches per plant. Tiller / branches per plant can be manipulated through fertilizer application provided there are a reasonable number of plants for branching or tillering.

Lodging: Increased crop stand causes plants and stem to become smaller, weaker and often taller to premature lodging of the crop. Cereals and millets are more susceptible to lodging at high crop densities. Lodging after flowering leads to heavy losses in cereals.

Environmental factors influencing crop density

Major environmental factors influencing crop density include solar radiation, rainfall / irrigation temperature, wild, relative humidity and management factor like nutrient management.

Plant density and yield relationships

Optimum plant population is very important to achieve higher yield. Crop density per unit area should be such that all plants get equal opportunity for the resources for growth and development. Summation of yield from such plants will result in maximum yield per unit area.

Crop geometry: Arrangement of plants in the field denotes planting geometry or planting pattern. It influences the crop yield through its influence on light interception, rooting and nutrient - moisture extraction pattern.

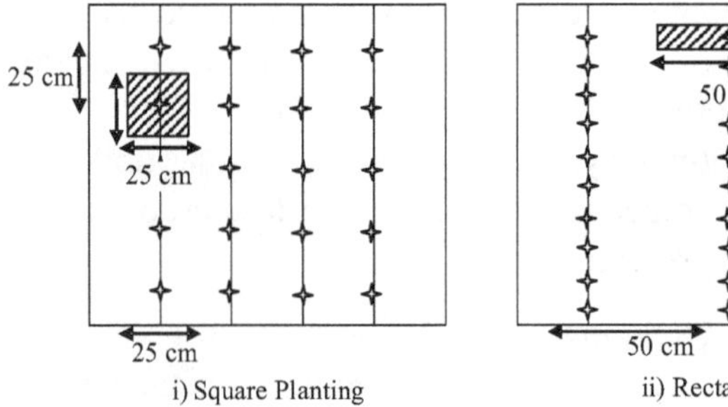

i) Square Planting ii) Rectangular Planting

i) **Square planting:** Here the inter (between the rows) and intra (between the plants) row distance are equal. It is reasonable to expect that square arranged plants will be more efficient in utilizing light, water and nutrients available to individual plant than rectangular arrangement. Square planting is advantageous in planting and intercultivation in both the direction.

ii) **Rectangular planting:** Sowing of crop with seed drill is a standard practice. Wider inter rows with closer intra row is the rectangular planting and is very common in most of the crops. This arrangement is adopted mainly to facilitate intercultivation especially in low volume crops.

iii) **Miscellaneous planting arrangements:**

a) **Random planting:** Plants are transplanted or seeds are broadcasted randomly without definite row arrangement.

b) **Skip row planting:** Skipping of every alternate rows and the population is adjusted by reducing intra row spacing.

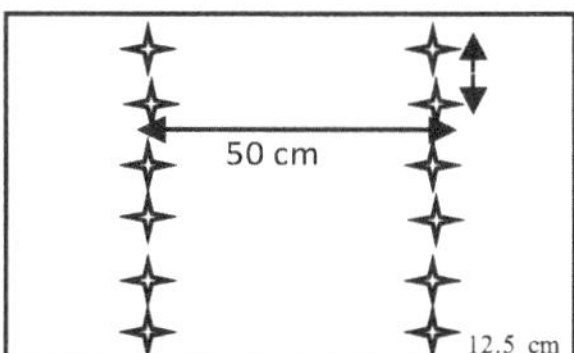

c) **Paired row planting:** It is also a skip row planting where in the plant population is adjusted by reducing inter row spacing of alternate rows. It is adopted specially for intercropping and irrigation under scarce water situation and mechanical harvesting in sugarcane.

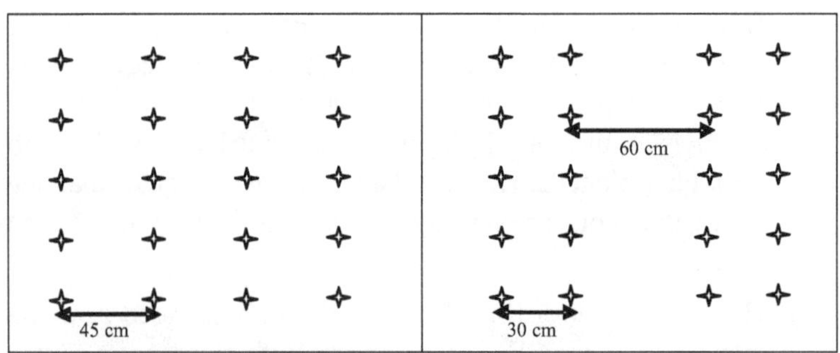

a) Regular planting b) Paired row planting

$$\text{Seed rate (kg ha}^{-1}\text{)} = \frac{100 \times T}{P \times R} \times \frac{100}{pp \times g} \quad \text{or} \quad \frac{P \times T}{pp \times g}$$

Where, T = Test weight (g)

P = Intra row spacing (cm)

R = Inter row spacing (cm)

pp = Purity percentage (%)

Where, P = No. of plants per sq. m

T = Test weight (g)

pp = Purity percentage (%)

g = Germination percentage

Chapter 11

Crop Nutrition-Manures, Fertilizers and Nutrient Use Efficiency

Crop nutrition / plant nutrition refers to the inter relationships of mineral elements in the soil and their role in plant growth and development. This inter relationship involves a complex balance of mineral elements essential and beneficial for optimum plant growth.

Plants require certain element for their growth and development. There are about 60 chemical elements present in the plant and it doesn't mean that all are essential for normal growth and development.

Arnon and Stout (1939) proposed criterion of essentiality. Which were refined by Arnon (1954) and are popularly known as **Arnon's criteria of essentiality.** They are :

1. The deficiency of element makes it impossible for the plant to complete the vegetative or reproductive stage of its life.
2. The deficiency of element in question can be corrected only by supplying that particular element.
3. The element is directly involved in nutrition and metabolism of the plant quite apart from its possible effect in correcting some micro-biological or chemical conditions in the soil.

Based on the above criterion, 16 elements are considered essential for plant growth and development. They are classified as :

I. Based on quantity of nutrients present in plants are grouped into

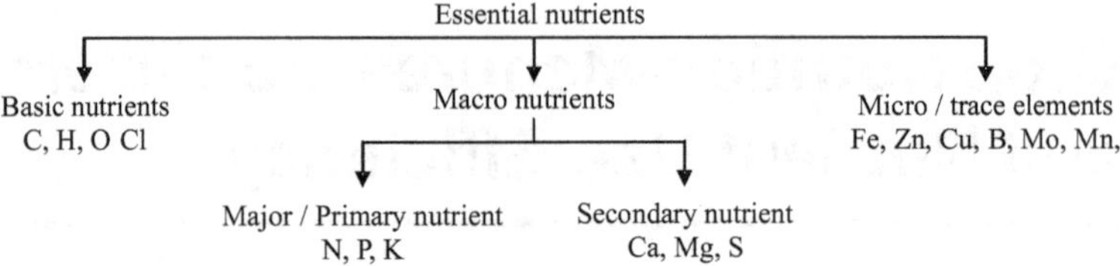

Basic nutrients: Contribute 96 per cent of the total dry matter of plants and are taken in the form of water and air (CO_2).

Macro nutrients: Are required in large quantity (> 1 ppm) and are further divided into :

a) **Major / Primary nutrients:** Are required in bulk and applied externally Ex. N, P, K.

b) **Secondary nutrients:** They are called secondary because they applied (unintentionally or accidentally) inadvertently to the soil while applying N, P, K fertilizers Ex. Ca, Mg, S.

Micro nutrients: They are also called trace elements and are required in minute quantity (<1 ppm). Even a slight deficiency or excess concentration of these elements is harmful to the plant.

II. Based on their function in plants

i) Elements that provide basic structure to the plant: C, H, O.

ii) Elements useful in energy storage, transfer, bonding: N, P, S.

iii) Elements necessary for charge balance / translocation: K, Ca, Mg.

iv) Elements involved in enzyme activation & electron transfer : Fe, Mn, Zn, Cu, B, Mo, Cl.

III. Based on mobility

A) **Mobility in the soil:** Knowledge of mobility in soil helps to derive the method of application of these nutrients. Based on mobility, the nutrients are classified as :

i) **Mobile:** These are highly soluble and are not adsorbed on clay complex. Ex : NO_3^-, SO_4^{2-}, BO_3^{2-}, Cl^-, Mn^{2+}.

ii) **Less mobile:** They are also soluble but are adsorbed on clay complex. Ex : NH_4^+, K^+, Ca^{2+}, Mg^{2+}, Cu^{2+}.

iii) **Immobile:** They are highly reactive and get fixed in the soil. The plant roots as to go in search of these nutrients for absorption. Ex: $H_2PO_4^-$, HPO_4^{2-}, Zn^{2+}.

B) **Mobility in Plants:** Knowledge of mobility in plant helps to find out the deficient nutrient. A mobile element in plant moves to growing point and express the deficiency in lower leaves and immobile elements shows deficiency in top leaves. Based on these nutrients are grouped as :

i) **Highly mobile:** N, P, K, Mg

ii) **Moderately mobile:** Zn

iii) **Less mobile:** S, Fe, Mn, Cu, Mo, Cl

iv) **Immobile:** Ca, B

Forms of nutrient uptake by plants

Nutrient element	Form of uptake	Nutrient element	Form of uptake
Carbon	CO_2	Iron	Fe^{2+}, Fe^{3+}
Hydrogen	H_2O	Manganese	Mn^{2+}
Oxygen	CO_2, H_2O	Zinc	Zn^{2+}
Nitrogen	NH_4^+, NO_3^-	Copper	Cu^{2+}
Phosphorus	$H_2PO_4^-$, HPO_4^{2-}	Boron	$B_4O_7^{2-}$, $H_2BO_3^-$
Potassium	K^+		& HBO_3^{2-}
Calcium	Ca^{2+}	Molybdenum	MoO_4^{2-}
Magnesium	Mg^{2+}	Chlorine	Cl^-
Sulphur	SO_4^{2-}		

Factors influencing nutrient availability: Important factors influencing nutrient availability in the soil are :

a) Native soil fertility or nutrient content of soil
b) Soil reaction (pH)
c) Relative activity of micro-organisms which play vital role in nutrient release through mineralization
d) External addition of manures and fertilizers
e) Edaphic / soil factors *viz.*, soil temperature, moisture, aeration *etc.*

Mechanism of nutrient movement in soil

Nutrients absorbed by the plant through roots along with the water. Nutrients move to the roots through 3 ways :

1. **Mass flow:** Plant nutrients move along with the movement of water (Soil solution). This is the major type accounts to 80 per cent of total.
2. **Diffusion:** This is the normal dispersion of nutrients from the region of their higher concentration to lower. When roots absorb nutrients beneath that develop region of lower concentration and nutrient movement occur.
3. **Root interception:** Is the extension of plant roots into new soil areas where they are untrapped.

Factors affecting nutrient absorption: These factors are grouped as:

A) External factors	B) Internal factors
i) Oxidation– reduction state of elements	i) Cell wall
ii) Concentration of elements	ii) Type of cell
iii) Moisture content of soil	iii) Stage of development
iv) Aeration – root respiration & energy release	iv) Transpiration rate
v) Temperature	
vi) Soil pH	

Manures and fertilizers : Plant nutrients are lost from the soil in different ways *viz.,* removal by the crop, weeds, lost through leaching, erosion, volatilization and denitrification. The importance of manures and fertilizers for increasing agricultural production to meet the challenges for food, fiber, forage, fuel are increasing because of progressive expansion in the nutrient deficiencies in India.

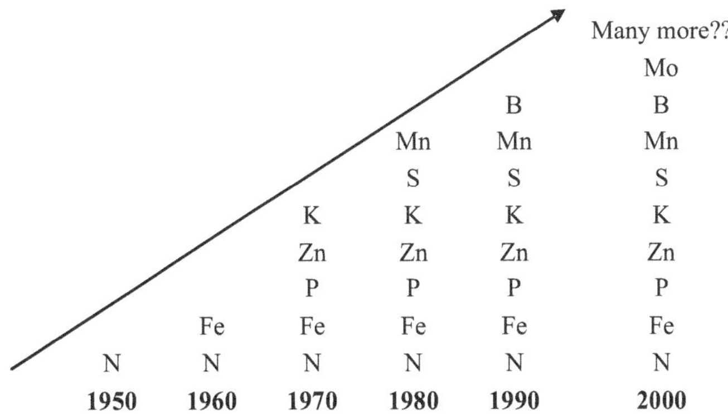

Diagrammatic representation of occurrence of soil nutrient deficiency in India

When crop requirements are higher than soil supplying power, nutrients are supplied through manures and fertilizers.

Manures: Manures are plant and animal wastes that are used as sources of plant nutrients. They release nutrients after their decomposition. Manures are grouped into **Bulky & Concentrated organic manures** based on the nutrient content.

Bulky organic manures: These contain small percentage of nutrients and are applied in large quantities. Hence they are called bulky. Farm yard manure (FYM), Compost and green manures are most widely used bulky organic manures. Use of these manures has several advantages like:

1. Supply of all plant nutrients including micro-nutrients.
2. Improve soil physical properties like structure, WHC *etc.*

3. Serves as source of food for soil organisms (Micro organisms).
4. Improves soil chemical properties like soil reaction, electrical conductivity, *etc.*

1. **Farm yard manure (FYM):** FYM refers to the decomposed mixture of dung and urine of farm animals along with litter and left over material from roughages or fodder fed to the cattle. On an average, well decomposed FYM contain 0.5 per cent N, 0.2 per cent P_2O_5 and 0.5 per cent K_2O.

Factors influencing composition of FYM

The quality and general characteristics of FYM are influenced by:

i) **Source of manure:** The composition of manure varies with the kind of animal producing it. Sheep, goat, and poultry manures are rich in N, P_2O_5 and K_2O than Cow, horse and pig.

NPK content of fresh excreta of farm animals

Animal	Dung	Urine
	$N : P_2O_5 : K_2O$	$N : P_2O_5 : K_2O$
1. Cows & Bullocks	0.4 : 0.2 : 0.10	1.0 : 0.01 : 1.35
2. Sheep & Goat	0.75 : 0.5 : 0.45	1.35 : 0.5 : 2.1
3. Horse	0.55 : 0.3 : 0.4	1.35 : Trace : 1.25
4. Pigs	0.55 : 0.5 :0.4	0.4 :0.1 : 0.45

ii) **Food of the animal:** This is one of the most important character determining the quality of manure. If the food source is rich (Pulse), then the nutrient in the excreta will also be rich. On an average, 70-90 % N & P and 90-99 % K and 50 % Organic matter in the food consumed will appear in dung or urine.

iii) **Age and Condition of animal:** Young and healthy animals retain large percentage of nutrients for their growth than that of mature and sick animals. Hence, dung and urine of young & healthy animal has lower nutrient than mature & sick.

iv) **Function of the animal:** Animals producing milk and wool absorb more nutrients than the animals which are in rest.

v) **Manner of storage:** Bad storage leads to loss of nutrients through volatilization & leaching.

vi) **Nature of litter:** Composition of FYM varies with litter or bedding material.

The daily collection of dung and urine are placed in pit / trenches or heaped and allowed for decomposition. The loss in nutrients occurs both at handling and storage.

A) **Handling losses:** Handling loss may be loss of liquid (Urine) or solid (Dung).

i) **Loss of liquid portion or urine**: Failure to recover or retain the urine of farm animals results in serious loss. It may be due to uncemented floor of cattle shed. Large quantity of N lost through volatilization as below :

$$\underset{\text{Urea in Urine}}{CO(NH_2)_2 + 2\,H_2O} \longrightarrow \underset{\text{Ammonium Carbonate}}{(NH_4)_2CO_3}$$

$$(NH_4)_2CO_3 + 2\,H_2O \longrightarrow 2\,NH_4OH + H_2CO_3$$

$$NH_4OH \longrightarrow NH_3\uparrow + H_2O$$

ii) **Loss of solid portion or dung:** Loss of dung is because of its use as fuel and no effort to collect when animal goes out for grazing.

B) **Storage losses:** Normally farmers heap their daily collection at one place either in trench or above ground surface for months together. During this period, the manure remains exposed to the sun & rains and nutrients are subjected for the following losses.

i) **Leaching:** About half of N & P and 90 per cent of K present in FYM are water soluble. When the manure is exposed to leaching action of rain water, the nutrients are liable to get washed off.

ii) **Volatilization:** During storage, ammonia produced by decomposition of urea and other nitrogenous compounds of urine may be evaporated.

Partially rotten FYM has to be applied three to four weeks before sowing while, well rotten manure can be applied immediately before sowing. The existing practice of leaving manure in small heaps scattered in the field for long period leads to loss of nutrients. These losses can be reduced by spreading the manure and incorporating by ploughing immediately after application.

The entire quantities of nutrient present in FYM are not available immediately. About, 30 per cent N, 60-70 per cent P_2O_5, and 70 per cent K_2O are available for the first crop. In general, about 40 per cent of total nutrients are available for first crop, 30 per cent for second, 20 per cent for third crop and 10 per cent for succeeding crops.

Chemical preservatives are added to the FYM to decrease N losses. The commonly used preservatives are a) Gypsum and b) Super phosphate.

Super phosphate has been extensively used as a manure preservative. Since ordinary super phosphate contains 50-60 per cent gypsum besides mono calcium phosphate, the action of super phosphate is similar to that of gypsum.

Further, super phosphate preservative has additional advantage *viz.*,

a) Reduce the loss of nitrogen as ammonium.

b) Increase phosphorus content of manure (enrichment).

c) Tri calcium phosphate produced is acid soluble which increases phosphorus use efficiency under acidic condition. It is recommended to apply 0.5-1 kg SSP / day / animal.

Compost: Compost is a mass of well rotten organic matter of any wastes. Composting is a biological process in which aerobic and anaerobic micro-organisms decompose the organic matter and reduce C : N ratio of the refuse.

The compost made from farm wastes like sugarcane trash, paddy straw, weeds, other plants and wastes is called farm compost. On an average, it contain 0.5 : 0.15 : 0.5 % NPK.

The compost made from town refuses like night soil, street sweepings and dustbin refuses is called town compost. On an average it contain 1.4 : 1.0 : 1.4 % N, P_2O_5, K_2O.

Night soil: Night soil is the human excreta, both solid and liquid. It is rich in NPK than FYM & Compost. On an average night soil contain 5.5 : 4.0 : 2.0 % N, P_2O_5, K_2O.

Sewage and sludge: In the modern system of sanitation adopted in cities and towns, human excreta are flushed out with water which is called as sewage. The solid portion of sewage is called sludge and liquid portion is sewage water.

Vermi compost: Compost that is prepared with the help of earth worms is called vermicompost. Earth worm consume large quantities of organic matter and excrete soil as cast. The cast of earth worms have several enzymes and are rich in plant nutrients, beneficial bacteria and mycorrhizae. On an average, it contains 3 % N, 1 % P_2O_5 and 1.5 % K_2O. The process of preparing compost using earth worms is called as vermicomposting.

Sheep and goat manure: manure prepared with the droppings of sheep and goats, which contain higher nutrients than FYM & Compost. On an average, it contains 3 % N, 1 % P_2O_5 and 2 % K_2O. Sheep and goat manure is applied in two ways like :

a) Composting:

b) Sheep penning: They left in the field overnight and the dropping (Fecal matter & Urine) are incorporated with harrows / cultivators.

Poultry manure: The excreta of birds ferment very quickly. If it is left exposed for 30 days, 50 per cent N will be lost. On an average, it contain 3 % N, 2.63 % P_2O_5 and 1.4 % K_2O.

Green manures: The practice of incorporating or turning into the soil undecomposed green plant materials for the purpose of improving physical condition as well as fertility.

Crop residues: Residues left after the harvest of the economic portions are called crop residues / straw. These crop residues can be recycled by the way of incorporation, composting or mulch material.

Agroindustrial wastes: Wastes of agro based industries viz., sugar, rice mill etc also serve as manures.

a) **Rice husk:** It is the major byproduct of rice mill. Unhulled paddy contains 20-25 % husk. It may be used as bedding material in cattle shed or incorporated into wet soil. It contain 0.3-0.4 : 0.2-0.3 : 0.3-0.5 % NPK.

b) **Bagasse:** Byproduct obtained during the process of sugar manufacturing. It contain 0.25 % N, 0.12 % P_2O_5.

c) **Pressmud:** Byproduct obtained during the process of sugar manufacturing. It contains 1.25 % N, 2 % P_2O_5 and 20-25 % organic matter.

Others like tea waste, coffee pulp, coir waste etc have manorial value.

Concentrated organic manure

They have higher nutrient content than bulky organic manures. The important concentrated organic manures are oil cakes, blood meal, fish meal, meat meal, fish manure etc. Oil cakes are dried solid portion remaining after extraction of oils. These are of two types.

a) **Edible oil cakes**: Which can be safely fed to the livestocks. Ex. Groundnut and coconut cake.

b) **Non-edible oil cakes:** Which are not fit for feeding livestock Ex: Castor, Neem, Hippe *etc.*

Oilcakes have to be well powdered before application for even distribution and quicker decomposition. The average nutrient content of different oil cakes are :

Oil cake	Nutrient content (%)		
	N	P	K
Non edible			
1. Castor cake	4.3	1.8	1.3
2. Pongamia cake	3.9	0.9	1.2
3. Mahua / Hippe cake	2.5	0.8	1.2
Edible oil cakes			
4. Coconut cake	3.0	1.9	1.8
5. Groundnut cake	1.3	1.5	1.3
6. Linseed cake	4.9	1.4	1.3
7. Sesamum cake	6.2	2.0	1.2

The blood of slaughtered animal and meat is dried and powdered to use as manure which contain

Organic manure	Nutrient content (%)		
	N	P	K
1. Blood meal	10 -12	1 – 2	1.0
2. Meat meal	10.5	2.5	0.5
3. Fish meal	4 – 10	3 - 9	0.3 – 1.5
4. Horn & Hoof meal	13	-	-

Bio-fertilizers: The term "Bio-fertilizer" refers to selective micro-organisms like bacteria, fungi and algae which are capable of fixing atmospheric nitrogen or convert insoluble phosphorus in the soil to available forms. Bio-fertilizers are cost effective, eco-friendly and renewable sources of plant nutrients to supply plant nutrients.

Based on the type of micro-organisms, bio-fertilizers are classified as:

i) Bacterial biofertilizers: Ex. Rhizobium, Azospirullum, Azatobacter, Phospho bacteria.

ii) Fungal biofertilizers: Ex. Mycorrhizae.

iii) Algal biofertilizers: Ex. Blue Green Algae (BGA), Azolla.

iv) Actinomycetes: Ex. Frankia.

Based on the association, bio-fertilizers are grouped as

i) Free living : Ex. Azotobacter

ii) Symbiotic : Ex. Rhizobium

Bio-fertilizers are mostly multiplied in the laboratory. However, BGA, Azolla can be mass multiplied in the field. Bio-fertilizers are used either as seed treatment or seedling root dip or soil application.

Fertilizers: Fertilizers are inorganic / synthetic substances containing one or more plant nutrients in easily soluble and quickly available form. Because of their availability in concentrated form, they have the advantage of being small quantity for storage, handling. In addition, crop may be supplied with exact quantity in the time required for their growth.

Fertilizers are classified into

1. Basded on number of nutrients present are classified as:

a) **Straight fertilizers:** Are those which supply only one primary plant nutrient Ex. Urea (46 % N), SSP (16 % P_2O_5) and MOP (60 % K_2O).

b) **Complex fertilizers:** Contain two or more primary nutrients in chemical combination. Ex. Di Ammonium Phosphate (DAP – 18 % N & 46 % P_2O_5), 17-17-17, 15-15-15 *etc.*

c) **Mixed fertilizers:** Are physical mixtures of straight fertilizers with definite proportion.

2. Based on the concentration of nutrients, fertilizers are classified as :

a) **Low analysis fertilizers:** Containing less than 25 % of primary plant nutrients. Ex. SSP (16 % P_2O_5), Sodium nitrate (16 % N).

b) **High analysis fertilizers:** Containing more than 25 % primary plant nutrient Ex. Urea (46 % N), DAP (18 % N and 46 % P_2O_5), Anhydrous ammonia (82 % N).

3. Based on the physical form are classified as:

a) **Solid form:** They may be powder (SSP), Crystal (Amm. Sulphate) or Prills (Urea super granules).

b) **Liquid form:** Prepared for applying with irrigation water, herbicides / pesticides etc. They may be (i) Clear liquid fertilizer *i.e.*, Completely dissolved (ii) Suspension liquid fertilizers *i.e.*, Suspended as fine particles.

4. Acid forming Fertiliges: Application of fertilizers increases acidity or basicity of soils depending on the nature of fertilizers.

1. Fertilizers which leave an acid residue in the soil are called **acid forming fertilizers**. The amount of calcium carbonate required to neutralize the acid residue is referred as **equivalent acidity.**

Acid forming fertilizers:

Fertilizers	Acid equivalent
Ammonium Chloride	128
Ammonium Sulphate	110
Amm. Sulphate Nitrate	93
Ammonium Phosphate	86
Urea	80

2. Fertilizers which leave alkaline residue in the soils are called **alkaline or base forming fertilizers**. The amount of calcium sulphate required to neutralize basic or alkaline residue is referred as **equivalent basicity.**

Base forming fertilizers:

Fertilizers	Base equivalent
Sodium Nitrate	29
Calcium Nitrate	21
Calcium Cynaide	63
Di Calcium Phosphate	25

Fertilizer Grade: Fertilizer grade refers to the guaranteed minimum percentage of N, P_2O_5, K_2O contained in a fertilizer material. Ex. 28-28-0, 20-20-0, 14-35-14, 17-17-17, 14-28-14.

Nitrogenous Fertilizers: The fertilizers which contain nitrogen are called as nitrogenous fertilizers. Based on the chemical form, nitrogen fertilizers are classified as :

1. **Nitrate fertilizers:** These are having nitrogen in the form of nitrate (NO_3). Ex. Sodium Nitrate ($NaNO_3$) – 16 % N, Calcium Nitrate ($Ca(NO_3)_2$) – 15.5 % N.
2. **Ammonium Fertilizers:** These are having nitrogen in the form of ammonia (NH_4^+).

   Ex. Ammonium sulphate [$(NH_4)_2SO_4$] – 20 % N,

   Ammonium chloride (NH_4Cl) - 24-26 % N,

   Ammonium phosphate [$NH_4H_2PO_4$) – 20% N & 20% P_2O_5,

   Anhydrous ammonia (NH_3) – 82 % N

   Ammonia solution - 20-25 % N.
3. **Nitrate & Ammonium fertilizers:** They have both nitrate & ammonium

   Ex. Ammonium nitrate [NH_4NO_3] – 33-34 % N,

   Calcium ammonium Nitrate [$CaNH_4NO_3$) – 20 % N,

   Ammonium sulphate nitrate ASN - 26 % N,
4. **Amide form**

   Ex. Urea [$CO(NH_2)_2$] – 46 % N,

   Calcium Cynamide $CaCN_2$ – 21% N,

General characteristics of nitrogen fertilizers

1. **Nitrate fertilizers:** NO_3^- form is readily soluble in water and more readily utilized by the plants. This is highly mobile form and subjected to leaching and can not be fixed to the clay colloids due to their negative charge.

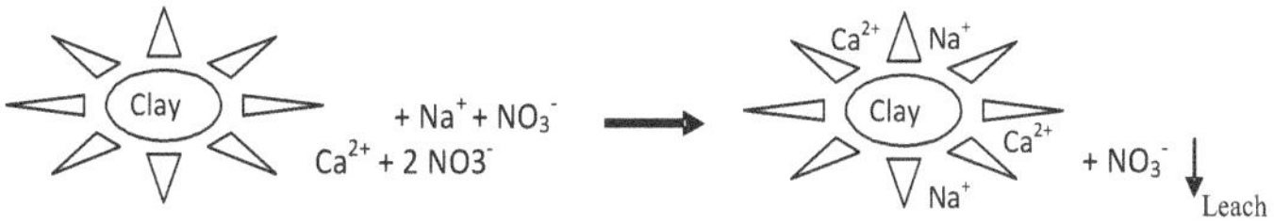

Clay $Ca^{2+} + 2 NO3^-$ $+ Na^+ + NO_3^-$ → Clay (Ca^{2+}, Na^+, Ca^{2+}, Na^+) $+ NO_3^-$ ↓ Leach

Since the nitrate fertilizers are the salts of Na, K & Ca, they are alkaline / basic and can be applied to acidic soils.

2. **Ammonical fertilizers:**

   a) They are also readily soluble in water.

   b) Slowly utilized by the plant as they need to convert to nitrate form

   $NH_4 \xrightarrow{\textit{Nitrosomanas}} NO_2 \xrightarrow{\textit{Nitrobacter}} NO_3$

   However, certain crops like rice can directly use ammonia form also:

   c) They are resistant to leaching losses as they are readily absorbed on the colloidal complex

   d) All Ammonical fertilizers are acidic in nature and can be used in alkaline soils.

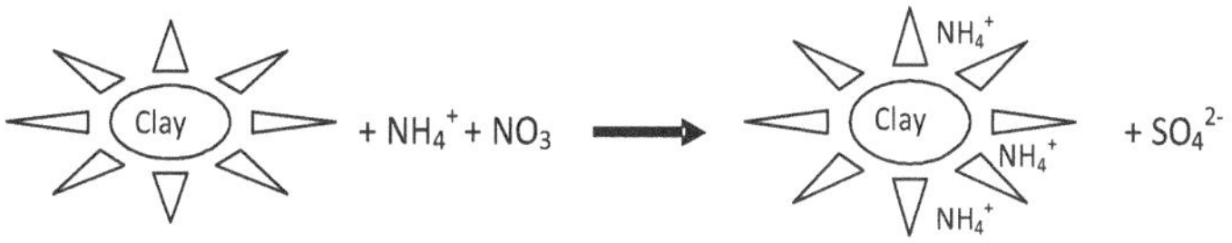

Clay $+ NH_4^+ + NO_3$ → Clay (NH_4^+, NH_4^+, NH_4^+) $+ SO_4^{2-}$

3. **Nitrate and ammonical fertilizers:** They have the characteristics of both Ammonical and nitrate fertilizers and are slightly acidic in nature.

4. **Amide fertilizers:**

   a) They are carbon compounds and are technically called as **organic fertilizers.**

   b) Readily soluble in water and easily decomposed by micro-organisms in the soil.

   c) Less acidic than ammonical fertilizers

$$CO(NH_2)_2 + 2\ H_2O \rightarrow (NH_4)_2CO_3$$
$$(NH_4)_2CO_3\ + 3\ O_2\ \rightarrow 2\ HNO_2\ + 3H_2O\ + CO_2$$
$$2\ HNO_2 + O_2 \rightarrow 2\ HNO_3$$

Slow releasing Nitrogen fertilizers: Rate of release / solubility of nitrogen fertilizer can be reduced by:

a) **Use of substances having low solubility:** Developed by condensation products of urea with formaldehyde (35 % N), Crotonylidene di urea (CDU-30 % N) and Iso butylidene di urea (IBDU – 30 % N).

b) **Physical barriers:** Nitrogenous fertilizers are coated with substances that reduce their solubility. Ex. Sulphur coated urea (SCU – 36-38 % N) – S barrier around.

c) **Modifying the size of urea:** Ex. Urea super granules - big size (1-3g) – slowly dissolves.

d) **Nitrification inhibitors:** Nloss from applied fertilizers takes after conversion to nitrate form by nitrification process with the activity of micro-organisms.

$$NH_4 \xrightarrow{\textit{Nitrosomanas}} NO_2 \xrightarrow{\textit{Nitrobacter}} NO_3$$

Several chemicals are toxic to nitrifying bacteria, when these are added to the soil, they temporarily inhibit the nitrification process. Ex. N – Serve, AM (2-amino 4-Chloro 6-methyl pyrimidine). These are however expensive and not cost effective.

Hence, some of the indigenous materials like neem, pongamia oil can also be used.

Ex. Neem coated urea, Neem blemded urea *etc.*

AM (2-amino 4-Chloro 6-methyl pyrimidine.

N-Serve (2-Chloro, 6 trichloro methyl pyridine)

Phosphatic fertilizers

The phosphorous content of fertilizers are expressed in oxidized form (P_2O_5) and are converted with the following factor.

% P = % P_2O_5 X 0.43

% P_2O_5 = % P X 2.29

When P_2O_5 solubelize in water, which produce Meta phosphoric and ortho phosphoric acid.

Classification of Phosphatic fertilizers: Based on the solubility and availability to the crops, phosphatic fertilizers are classified as

1. **Water soluble phosphatic fertilizers**

   a) These are available in the form of mono calcium or ammonium phosphate.

   $Ca - H_2PO_4$ \
   H_2PO_4

   b) They are readily absorbed by the plant as they contain phosphorus in $H_2PO_4^-$ form.

   c) These are readily transformed in the soil into water insoluble form. Hence, the leaching losses are very minimum.

   d) They can be used for short duration crops Ex. Rice, Wheat, Ragi *etc.*

   e) As they contain phosphorus as Ca & calcium ammonium salts, their use is restricted to neutral and alkaline soil. While in acidic soil they fix as Fe and Al phosphates.

Ex. Single Super Phosphate (SSP – 16 % P_2O_5),

Double Super Phosphate (DSP – 32 % P2O5)

Tripple Super Phosphate (TSP – 48 % P_2O_5),

Mono Ammonium Phosphate (48 % P_2O_5 + 11 % N)

Di Ammonium Phosphate (DAP – 46 % P_2O_5 + 18 % N).

2. **Citric acid soluble phosphatic fertilizers**

   a) These contain soluble phosphoric acid or Di Calcium Phosphate.

   $Ca - HPO_4$ \
   |\
   $Ca - HPO_4$

b) Suitable for acidic soil as citrate soluble form will convert to mono calcium phosphate or water soluble form and there is less chance of getting fixed as Fe & Al Phosphates. Further, they are basic in reaction.

c) These fertilizers can be used for long duration crops like sugarcane, tea, coffee *etc.*

Ex: Basic slag (14-18 % P_2O_5)

Di Calcium Phosphate (35-40 % P_2O_5).

3. Water and Citrate Insoluble Phosphatic fertilizers

a) Contain Phosphoric acid which is insoluble in water and citric acid.

b) Suitable for strongly acidic soils.

c) Availability can be improved by ploughing along with green manures.

d) Suitable for Coffee, Rubber, Cocao in hilly tracts as the soils are acidic.

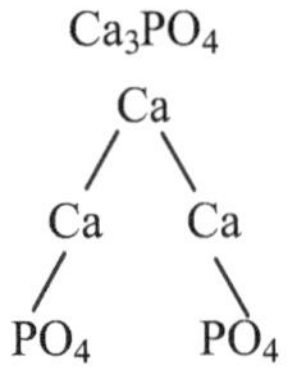

Ex. Rock Phosphate – 20-40 % P_2O_5.

Ca_3PO_4

```
        Ca
       /  \
     Ca    Ca
    /        \
  PO4        PO4
```

Potassic Fertilizers: Fertilizers containing potassium are grouped into 2 groups.

1. Fertilizer having K in Chloride form (KCl)

   Ex. Muriate of potash (MOP) – 58-60 % K_2O.

2. Fertilizer having K in non-chloride form (K_2SO_4)

   Ex. Sulphate of potash (SOP)–48 – 50 % K_2O

   Potassium nitrate (KNO_3) – 44 % K_2O + 13% N.

Potassium content in the fertilizer are always expressed as K_2O which can be converted with the following :

% K = % K_2O x 0.83

% K_2O = % K x 1.20

From the potassium nutrition point, all fertilizers are equally effective, but accompanying anions make some difference. In non-chloride form, sulphates and nitrates are also adds to the nutrition of plant but they are costly.

Potassium chloride (MOP) is suitable for most of the crops except sugarcane, sugar beet, potato and tobacco. In sugar crops, sugar accumulation is affected while in tobacco leaf, it reduces the burning quality in the presence of chloride and in potato chloride decreases for keeping and cooking quality.

Secondary nutrient fertilizers: Calcium, magnesium and sulphur are supplied to the plants incidentally along with NPK fertilizers and are not manufactured independently. Hence, they are considered as secondary.

The fertilizers containing secondary nutrients are:

Fertilizer	Nutrient content (%)			
	Calcium	Magnesium	Sulphur	Others
1. Calcium nitrate	19.4	-	-	-
2. Gypsum	29.2	-	18.6	-
3. Single super phosphate	19.5	-	12.5	16.0 P_2O_5
4. Epsom salt ($MgSO_4$)	-	9.6	13.0	-
5. Potassium sulphate	-	-	17.5	48-50 K_2O
6. Ammonium sulphate	-	-	24.2	21 N
7. Ammonium sulphate nitrate	-	-	12.1	26 N

Micro-nutrient fertilizers: These are required in minor quantity and are applied only in specific situation of their deficiency. Important micro-nutrient fertilizers are :

Source	Formulae	Nutrient content
1. Zinc sulphate	$ZnSO_4 \cdot 7H_2O$	22.35 % Zn
2. Ferrous sulphate	$FeSO_4 \cdot 7H_2O$	20.00 % Fe
3. Copper sulphate	$CuSO_4 \cdot 5H_2O$	25.35 % Cu
4. Borax or sodium borate	$Na_2B_4O_7 \cdot 10H_2O$	10.6 % B
5. Manganese sulphate	$MnSO_4 \cdot 4H_2O$	23.0 % Mn
6. Ammonium molybdate	$(NH_4)_6Mo_7O_{24} \cdot 4H_2O$	54.0 % Mo

Fertilizer Dose: Crop nutrient are met out by soil contribution and fertilizer application. Fertilizer recommendations are for :

a) Maintenance

b) Cation saturation ratio

c) Sufficiency

Maintenance concept implies the replacement of the exhausted or absorbed nutrient by the plant from soil.

Cation saturation ratio or Corrective concept refers that soil must have ideal ratio between exchangeable cation (65 % Ca, 10 % Mg, 5 % K and 20 % H).

Sufficiency level concept implies to the crop response or plant uptake excluding soil contribution.

Factors affecting fertilizer dosage: Several factors known to influence the dosage *viz.*

a) **Crop & Variety:** Quantity of fertilizer demanded is basically determined by the crop and variety. Explorative crops demand higher dosage and legumes can fix atmospheric nitrogen and demand nitrogen only till nodulation. Short duration and low yielding varieties demand lower dosage than long duration and high yielding variety / hybrids.

b) **Native soil fertility and other soil properties:** If the soil is natively fertile, the external dosage can be reduced as the maintenance dosage is marginal. Soil test ratings were established for organic carbon, NPK as follows:

Nutrient	Low	Medium	High
1. Organic Carbon (%)	< 0.5	0.5 – 0.75	> 0.75
2. Available Nitrogen (Kg ha^{-1})	< 280	280 – 560	> 560
3. Available P_2O_5 (Kg ha^{-1})	< 22.5	22.5-56.3	>56.3
4. Available K_2O (Kg ha^{-1})	< 141	141-336	> 336

If the soil are medium fertile, apply recommended dosage. If they are high, apply 25 per cent lower than recommended dose and if they are low, apply 25 per cent extra dose.

c) **Management:** If the farm is managed intensively without any limits for other factors, the optimum dosage boosts the yield. If there are limitations for any factor *viz.*, light, water, temperature *etc.*, the higher dosage is uneconomical.

Calculation of fertilizer dosage: Fertilizer recommendations are made in elemental or oxidized form of nutrient but not as such as fertilizers available. The amount of fertilizer to be applied can be calculated using the formulae :

Amount of fertilizer to be applied (Kg ha^{-1})

$$= \frac{\text{Recommended dosage (Kg ha}^{-1})}{\text{Nutrient content of fertilizer (\%)}} \times 100$$

Problem: Calculate the quantity of Urea, SSP, MOP required for paddy crop in 1 ha area with the recommendation of 100:50:50.

Solution: Data given - Recommended dosage: 100:50:50

Fertilizers - Urea (46% N), SSP (16% P_2O_5) and MOP (60% K_2O)

Calculation of Urea

$$\text{Amount of Urea required (Kg ha}^{-1}) = \frac{\text{Recommended dosage (Kg ha}^{-1})}{\text{Nutrient content of Urea (\%)}} \times 100$$

$$= \frac{100}{46} \times 100 = 217.39 \text{ Kg ha}^{-1}$$

Calculation of SSP

$$\text{Amount of SSP required (Kg ha}^{-1}) = \frac{\text{Recommended dosage (Kg ha}^{-1})}{\text{Nutrient content of SSP (\%)}} \times 100$$

$$= \frac{50}{16} \times 100 = 312.50 \text{ Kg ha}^{-1}$$

Calculation of MOP

$$\text{Amount of MOP required (Kg ha}^{-1}\text{)} = \frac{\text{Recommended dosage (Kg ha}^{-1}\text{)}}{\text{Nutrient content of MOP (\%)}}$$

$$= \frac{50}{60} \times 100 = 83.33 \text{ Kg ha}^{-1}$$

In situation of deviation in area the formulae can be modified as:

Amount of fertilizer to be applied

$$= \frac{\text{Area (ha) X Recommended dosage (Kg ha}^{-1}\text{)}}{\text{Nutrient content of fertilizer (\%)}} \times 100$$

Methods of fertilizer application: Appropriate method of fertilizer application is essential for:

a) Make the nutrients easily available for crop plant
b) Reduce fertilizer / nutrient loss
c) Ease of application

Suitable method for a particular situation depends on the nature of soil, crop and fertilizer material.

i) **Nature of soil:** Soil properties like texture, pH, CEC, nutrient and moisture status are important factors to be considered in selecting suitable method. If pH of soil is higher (>8), the ammonia volatilizes and it is not advisable to apply to apply ammonical fertilize. If the soil P content is less, P fertilizers as to be applied as band placement to reduce the fixation problems.
ii) **Nature of crop:** It depends on the root system and spacing. If the crop is closely spaced with shallow fibrous roots, fertilizer has to be applied in surface layer through broadcasting followed with irrigation. For widely spaced and deep rooted crop, deep point / band placement is advantageous.
iii) **Nature of fertilizer:** Fertilizer properties like form, solubility, mobility in soil decides the method of application. Granular /

powder form can be broadcasted, pellets can be point placed and liquid form can be fertigated.

Different methods of fertilizer application: Fertilizer may be applied either in solid or liquid form to the soil. The different methods of soil application of fertilizers are:

1. Application of fertilizers in solid form

a) **Broadcast:** The fertilizers is spread uniformly over the entire surface and then mixed with soil by tillage implements. Broadcasting of fertilizer is done at planting and as top-dressing.

b) **Placement:** Fertilizers are placed in the soil irrespective of the position of seed, seedling or growing plants before sowing or after sowing the crops.

i) **Plough sole placement:** Fertilizer is placed in a continuous band on the bottom of the furrow during ploughing. Each furrow covered as the next furrow is turned.

ii) **Deep placement of fertilizers:** Ammonia forming nitrogenous fertilizer are placed deep in the reduction zone in paddy fields to avoid losses. Deep placement of phosphatic fertilizers in deep black soils for paddy also increases its efficiency.

c) **Localized placement:** Fertilizer is applied into soil close to seed or plant.

i) **Contact placements or combined drilling or drill placement:** Only small quantity of fertilizer and seeds combined and drilled so that germination may not be adversely affected.

ii) **Band placement:** Fertilizer is placed in Bands. These bands may be continuous or discontinuous (close to the seed or transplanted plant). Hill placement or ring placement can be followed when plants are widely spaced particularly in square planting. Row placement can be followed for placing fertilizers one side or both sides of the row by hand or a seed drill.

iii) **Pellet application:** Nitrogenous fertilizer is applied in the form of pellets 3-5 cm deep between the rows of the paddy crop. The fertilizer is mixed with the soil in the ratio of 1:10 and made into dough. Small pellets of convenient size are then made and deposited in the soft mud of paddy fields. Application of urea through mud balls and paper packet is convenient for deep placement.

2. Application of fertilizers in liquid form

a) **Starter solutions:** Solutions of fertilizers generally consisting of N-P_2O_5-K_2O in the ratio of 1:2:1 and 1:1:2 are applied to young vegetable plants at the time of transplanting, in place of watering.

b) **Foliar application:** This refers to the application of dilute solutions of fertilizers like urea (1-5 %) directly on the foliage. Higher concentrations may cause scorching of leaves. When applied along with other spraying operations of pesticides it is less costly.

c) **Direct application to soil:** With the help of special equipment anhydrous ammonia (a liquid high pressure upto 200 pounds per square inch or more) and nitrogen solutions are directly applied to the soil. This practice is very popular in USA.

d) **Application through irrigation water (fertigation):** Straight and mixed fertilizers easily soluble in water are allowed to dissolve in the irrigation stream. The flow of water is regulated based on the dosage.

Fertilizer especially secondary and micro-nutrients are also applied to the plant as:

a) **Root dipping:** The roots of the seedlings are dipped in nutrient solution before transplanting.

b) **Foliar spray:** Fertilizers solutions are sprayed on the foliage of standing crop.

Time of Application: The time of fertilizer aimed at providing nutrient in sufficient quantities to meet the crop demand and avoiding excessive availability (toxicity) at all the stages and reducing losses. The time of application depends on crop uptake pattern, soil properties, nature of fertilizer material and utilization of carbohydrates.

a) **Crop uptake:** NPK are taken by the crop in large quantities in early stage in cereals. Legumes require nitrogen until root nodulation and P & K gradually throughout the crop growth. Further, duration and nature of crop also influence the time greatly.

b) **Soil properties:** Solubility and availability of nutrients depends on soil physical and chemical properties. In light textured soils, N fertilizer has to be applied in more number of splits.

c) **Nature of fertilizer:** 'N' fertilizer is soluble and highly mobile while, 'P' is subjected for fixation problem and become immobile. N fertilizer has to be applied in split dose while, P & K as basal dose.

d) **Utilization of carbohydrates:** The level of carbohydrates and nitrogen are inversely related. When large quantity of N fertilizer applied, the carbohydrate content reduces. The time of N application depends on the end product. In fodder crops, succulent leaves with higher protein are preferred. Hence, applications of N in several splits are essential. N shouldn't be applied in late stage where the end product is carbohydrates. In Sugarcane, sugar recovery reduces with increased N.

Time of fertilizer application can be divided into

a) **Basal application:** Application of fertilizer before or at the time of sowing is known as basal application. A portion of the recommended dose of N and entire dose of P & K are applied as basal in most of the crops.

b) **Split application:** Application of recommended dose of fertilizer in two or three splits during the crop period is known as split application. Application of fertilizer in standing crop is known as **Top dressing**.

Balanced fertilization: Refers to the application of NPK nutrients to soil in quantities to bring the balance in nutrients to meet the requirement of any a specific crop.

Integrated Nutrient Management (INM): Plant nutrients can be supplied through different sources *viz.*, organic manure, crop residues, bio-fertilizers, green manres, oilcakes, chemical fertilizers. For efficient utilization of nutrients and to produce crop with less cost, INM is the best approach.

INM is the judicious nutrient management system where in the nutrient demand of the crop is supplemented with 5 more sources considering the economics / cost.

Chapter 12

Growth and Development of Crops and Plant Ideotypes

Growth is defined as an irreversible change in the size of a cell, organ or whole organism. Commonly, growth is the increase in the amount of living material (protoplasm) which leads to an increase in cell size and ultimately cell division. Growth occurs only in living cells by metabolic process involved in the synthesis of proteins, nucleic acids, lipids and CHO at the expense of metabolic energy provided by photosynthesis and respiration.

Characteristics of growth

The main characteristics of growth are

1. Cellular growth
2. Cell division
3. Cell expansion
4. Cellular differentiation

Growth rates can be defined as increased growth per unit time and the rate of growth can be expressed mathematically.

Types of growth rate

1. Arithmetic growth rate
2. Geometric growth rate

Development is a term that includes all changes that an organism goes through during its lifecycle from germination of the seed to senescence.

Plant growth regulators

Plant growth regulators (phytohormons) are chemical substances influences the growth and differentiation of plant cells, tissues and organs. The plant growth regulators function as chemical messengers for intercellular communication. They work to ether coordinating the growth and development of cells.

Phytohormons

1. Auxins
2. Gibberlins
3. Cytokinies
4. Abiscic acid (ABA)
5. Ethylene

Classification of plant growth hormones (PGRs)

1. Growth promoter - Auxins, gibberellins and cytokinines.
2. Growth inhibitors - Abiscic acid (ABA) and Ethylene.

Vernalization : An artificial exposure of plants (or seeds) to low temperatures in order to stimulate flowering or to enhancing seed production. By satisfying the cold requirement of many temperate-zone plants, flowering can be induced to occur earlier than normal or in warm climates lacking the requisite seasonal chilling. Knowledge of this process has been used to eliminate the normal two-year growth cycle required of winter wheat by partially germinating the seed and then chilling it to 0° C (32° F) until spring it is possible to cause winter wheat to produce a crop in the same year.

Devernalization can be brought about by exposing previously vernalized plants or seeds to high temperatures, causing a reversion to the original non flowering condition. Onion sets that are commercially stored at near freezing temperatures to retard spoilage are thereby automatically vernalized and ready to flower as soon as they are planted. Exposure to temperatures above 26.7° C (80° F) for two to three weeks before planting, however, shifts the sets to the desired bulb-forming phase.

Phototropism: Is the response of plants to the relative lengths of light and dark periods. Based on this plants are divided into short day plants, long day plants and day neutral plant.

Plant ideotypes: Refers to plant type in which morphological and physiological characteristics are ideally suited to achieve high production potential and yield reliability. The concept of ideotype was given by **"Donald in 1968"** and he illustrated that there should be minimum competition between the crops and crop must be competent to compete with weeds. Ideotype is defined as **a biological model which is expected to perform or behave in a predictable manner within a defined environment.** On the basis of environment Donald and Hamblin (1976) identified two forms of ideotypes that is isolation ideotypes and competition ideotypes. Competition ideotypes are suitable for mixed cultivation.

The feature of crop ideotypes consists of several morphological and physiological traits which contribute for enhanced yield or higher yield than currently prevalent crop cultivars. The important features of ideotype for some crops are given below.

Wheat: According to Donald (1968), the ideotype for wheat crop has following features :

1. Short strong stem to avoid lodging.
2. Few small erect leaves to allow the sunshine into its canopy.
3. A large erect ear.
4. More number of fertile florets per unit area therefore more harvest index.
5. Presence of awns to increase the photosynthesis area.
6. A single culm to avoid wasteful vegetative growth.
7. Resistance to insect-pest and diseases.
8. Proper partitioning and translocation of assimilates.

Maize

1. Stiff, vertically oriented leaves above the ear (leaves below the ear should be horizontally oriented).
2. Maximum photosynthesis.
3. Efficient translocation of photosynthates to grain.
4. Long grain filling.
5. Slow leaf senescence.
6. Large cobs.
7. Cold tolerance in germinating seeds and young seedlings.
8. The height of the plant is to be 1.5 m in which cobs may be produced on the nodes near the tassel.

Cotton

1. Short and compact plant.
2. Determinate growth habit with uni model distribution of bolls.
3. Responsive to high fertilizer dose.
4. High degree of resistant to pest and diseases.
5. High physiological efficiency.

Rainfed upland rice

1. Short growth duration (85-100 days).
2. Effective deep root system.
3. Dwarf plant (<100 cm) with erect leaves and thick stem.
4. Early strong fertile tillering.
5. Synchronized flowering.
6. Good number of panicles (about 400/m^2).
7. Higher number of grains per panicle.
8. Moderate seed dormancy.
9. Resistance to insect-pest and diseases.

Ideotype for Dryland Farming

1. Short growth duration.
2. Effective root system.
3. Drought tolerance.
4. High yield potential with altered morphology *viz.*
   a) Plant with few leaves just sufficient to maintain photosynthetic output and growth (to minimize the use of water).
   b) Leaves horizontally disposed for better light interception contrary to vertically disposed under irrigated conditions.

Chapter 13

Cropping Systems and its Principles

Cropping systems are designed to mimic nature and bring diversity into our farming system. The cropping system should provide enough food for the family, fodder for cattle and generate sufficient cash income for domestic and cultivation expenses.

The objective of any cropping system is efficient utilization of all resources *viz.*, land, water and solar radiation, maintaining stability in production and obtaining higher net returns. The efficiency is measured by the quantity of produce obtained per unit resource used in a given time. The objective of traditional agriculture was to increase the production by two means:

a) By increasing area under cultivation.

b) By increasing the productivity per unit area of the crop.

But two more dimensions are added to modern agriculture

a) To increase the production per unit time.

b) To increase the production per unit space.

In the traditional cropping systems, mixtures and rotations were developed by the farmers over years of experience by trial and error to suit specific ecological and sociological conditions to attain yield stability, whereas modern scientific cropping has three pillars, *viz.,* 1. Genotype 2. Geometry of planting and 3. Management practices.

1. Genotype means genetic makeup of seed.
2. Geometry of planting means: Geometry of planting may be circular, rectangular, square type or cubical. It is indirectly related to plant population. Cubical pattern of planting has maximum plant population. Plant population may be defined as (i) size of area available to the individual plant, (ii) number of plants per unit area.
3. Management practices include all the practices of crop production. For the cropping system, management includes
   a) Type and arrangement of crops in time and space *i.e.* cropping pattern.
   b) Choice of variety.
   c) Method of stand establishment.
   d) Pest management and harvest.

What is a System: A system is a group of interacting components, operating together for a common purpose, capable of reacting as a whole to external stimuli: it is unaffected directly by its own outputs and has a specified boundary based on the inclusion of all significant feedbacks. For example, the human body is a system-it has a boundary (e.g., the skin) enclosing a number of components (heart, lungs) that interact (the heart pumps blood to the lungs) for a common purpose (to maintain and operating the living body).

Agro-ecosystem: Agro-ecosystems are ecological systems modified by human beings to produce food, fibre or other agricultural products. Like the ecological systems they replace, agro-ecosystems are structurally and dynamically complex. But their complexity arises from the interaction between socio-economic and ecological processes.

Cropping Systems: It is defined as the order in which the crops are cultivated on a piece of land over a fixed period or cropping system is the way in which different crops are grown. In the cropping systems, sometimes a number of crops are grown together or they are grown separately at short intervals in the same field.

Integrated farming systems: System of farming on a particular farm which includes crop production, raising livestock, poultry, fisheries, bee keeping, *etc.*, to sustain and satisfy as many needs of the farmers as possible. The objective of mixed farming is subsistence while higher profitability without altering ecological balance is important in farming systems.

Mono cropping: Growing same crop year after year on the same field. Ex: Rice-Rice-Rice.

Sole cropping / solid cropping: It is defined as the cultivation of one crop variety alone in the pure stand at normal density in a certain time and place.

Shifting cultivation: Forest land is cleared and cultivated for few years. Due to cultivation of the same crop on the same cleared forest land year after year soil productivity is lost and crop is shifted to other slashed and burnt area.

Livestock system: Land unit comprising of animals and auxiliary feed sources transforming plant biomass into animal products.

Farming system: Farming system is the scientific integration of different interdependent and interacting farm enterprises for the efficient use of land, labour and other resources of a farm family which provides year round income to the farmers specially located in the handicapped zone.

Crop rotation: Growing of different crops in succession on a piece of land in a certain sequence is called crop rotation. The rotation can include both cereals and legumes which are seeded for several years.

Ley farming: Ley farming is an system of farming in which grasses and legumes are included in proper rotation for hay, silage and pasture to cater the maximum livestock needs and to improve and conserve the soil fertility.

Alley cropping: Alley cropping is the cultivation of food, forage and other crops between rows of trees. It is a larger version of intercropping or companion planting conducted over a longer time scale. Alley cropping will provide profitable opportunities for farmers.

Multiple cropping: Growing two or more crops on the same field within a given year. It includes both sequential and inter cropping.

Mixed cropping: Two or more crops are mixed together in the same field at the same time without a definite row arrangement. Ex: cereals + legumes + millets + oilseeds.

Multistoried cropping: Growing more number of crops on the same field at the field to utilize vertical space more effectively. Ex: Coconut + Arecanut + pepper + Cocoa + Pineapple.

Basic principles of cropping system: The basic principles of cropping systems are as follows:

1. **Choose the crops that complement each other:** This involves choosing complementary crops and systems that share resources without causing nutrient deficiencies of neighboring or subsequent crops. Sowing nitrogen demand crops followed by nitrogen fixing legumes.
2. **Choose crops and cropping rotations which utilize available resources efficiently:** Examine the factors that limit crop productivity and design a rotation that emphasizes crops which best utilize available resources and minimize production risks. Strategies include choosing plants with different nutrient uptake rate, different heights for even distribution of sunlight, different rooting patterns, varying plant structures or different harvest times.
3. **Choose crops and cropping system that maintain and enhance soil fertility:** This includes that maintenance of nutrients such as nitrogen and carbon. Crop rotations with approximately 30-50 per cent nitrogen fixing leguminous and if

sufficient quantity of organic matter is added then 30 per cent N fixing legumes and if no organic manures are added then 50 per cent N fixing legumes should be included in the rotation. Again select the crops which produce large amount of organic matter in both above and below ground.

4. **Choose crops which have a diversity of growth cycles:** An ideal crop rotation would consists of early spring seeded, summer seeded, fall planted and perennial hay and pasture crops. However this may not possible if climatic conditions are unfavorable.

5. **Choose diverse species of crops:** Typically it is a good idea to rotate between grass and broad leaf for soil fertility restoration and for disease prevention. Changing the field each year between grass family plants and broadleaf plants helps to reduce the carry over pathogenic diseases organisms from year to year.

6. **Keep the soil covered:** Efforts should be made to grow sequences of crops that maximize solar radiation capture and minimize risks of soil erosion by maintaining the crops throughout the year so that rain drop impact on bare soil avoided.

7. **Strategically plan and modify your cropping system as needed:** The optimal crop rotation for in farm will evolve only through effective planning and many years of experimentation and observation, Diversified cropping systems will need more management, entire farm planning goals include house hold food security, income generation needs, livestock feed requirements, labour management, *etc.,* and also need to make adjustments for the prevailing weather and market conditions as the cropping year progress.

8. **Monitor the progress:** Make a plan and keep records. Learning from mistakes will result in more efficient crop production.

Benefits of cropping systems

1. **Maintain and enhance soil fertility:** Some crops are soil exhausting while others help to restore soil fertility. However, a diversity of crops will maintain soil fertility and keep production level high.
2. **Enhances crop growth:** Crops may provide mutual benefits to each others. For example, reducing lodging, improving nutrient availability though nitrogen fixation or even acting as wind breaks to improve growth.
3. **Minimize spread of diseases:** The more diverse the species of plants and the longer the period before the soil is resseded with same crop, the more likely disease problems will be avoided.
4. **Control of weeds:** Crops planted at different times of the year have different weed species associated with them. Rotating crops helps to prevent build up of any one serious weed species. The more different growth cycles the crops in rotation, fewer weeds will be able to adapt to the field conditions.
5. **Use of resources more efficiently:** Having diverse group of crops helps to more efficiently use the available resources like soil, nutrients, sunlight and water again minimizing the risk of nutrient deficiencies and drought. Apart labour, animal draft power and machinery are also utilized more efficiently as the time and harvest of crops spread out over the entire year.
6. **Reduces risk of crop failure:** Because of diverse group of crops help in preventing total crop failure as climate.
7. **Improved food and financial security:** Choosing an appropriate and diverse number of crops will lead to a more regular food production throughout the year. With lower risk of crop failure leads to greater reliability on food production and income generation.

Chapter 14

Crop Adaptation and Distribution

Crop adaption is closely related to genetics and plant physiology. Knowledge of basic of biological evolution and genetics is essential for an understanding of crop adaptation. Natural selection, the fundamental basis of adaptation has been atmost importance in determining the characteristic of present day crops and its still of great significance in modern plant breeding.

Ecology and ecosystem: Ecology and ecosystem influences adaption and distribution of crops. As such basic knowledge of ecology and ecosystem is necessary to understand adaptation and distribution of crops for maximum economic yields through efficient use of natural and applied inputs.

Ecosystem ecology is the integrated study of living and non living components of ecosystems and their interactions within an ecosystem frame work. This science examines how ecosystems work and relates this to their components such as chemicals, bedrock, soil, plant and animals.

Ecology is the scientific study of relations that living organisms have with respect to each other and their natural environment. Under ecology we study various ecosystems which are part of the biosphere. An ecosystem includes all the organisms and the non living environment that are found in particular place.

Types of Ecosystem

There are five types of ecosystem:

1. Forest ecosystem
2. Grass land ecosystem

3. Crop ecosystem
4. Desert ecosystem
5. Aquatic ecosystem

Ecopysiology is a branch of plant physiology which delas with the ecological control on growth, reproduction, survival, abundance and geographical distribution of plants.

Crop adaptation is special features that allow a plant or animal to live in a particular place of habitat. Soil is the store house for almost all the mineral for the growth of organisms. It also provides space for birth, growth and death of the organisms. Climate and vegetation of that place are responsible for variation in the physical and chemical properties of the soil. These characteristics of soil greatly influence the germination, growth behaviour and distribution pattern of plants.

Plants growing in different types of soil exhibit variations in their physiological and morphological features :

Oxylophytes	-	Plants growing in acid soils
Halophytes	-	Plants growing in saline soils
Psammophytes	-	Plants growing in sand
Lithophytes	-	Plants growing on rocks
Chasmophytes	-	Plants growing on rock crevices

Precipitation is one of the most important factors influencing the vegetation of a place. Most of the crop receives their water supply form rainwater. The amount of rainfall in different localities largely determines the nature of vegetation therein.

Based on rainfall the world vegetation can be classified as follows:

Rainfall (mm)	Vegetation
1. 0 to 13.24	Desert
2. 13.5 to 35.1	Semi arid grassland
3. 35.2 to 63.5	Dry sub tropical grass land savanna and open wood land
4. 63.6 to 114.3	Humid sub tropical forest
5. 114.4 to 203.2	Tropical rain forest

Based on water requirement plants are divided into

1. Hydrophytes
2. Mesophytes
3. Xerophytes

Based on temperature plants are divided into megatherm and microtherms in equatorial, tropical, temperate and arctic and alpine.

Climate and crop distribution: Basic principles and concepts relating to natural distribution of plants are useful in studying the plant adaptations. The plant geographer mainly studies the spatial relationship of plants both in past and present. He attempts to explain origin, development and distribution of plants.

The plant geographer Good (1953) formulated the basic principles of plant distribution and they are

i) Plant distribution is primarily controlled by the distribution of climatic conditions like light, temperature, moisture, wind *etc.*

ii) Plant distribution is secondarily controlled by distribution of edaphic factors like soil, parent material, physiography *etc.* these factors considered secondary because they are greately influenced by climate.

iii) Great movement of flora have been taken place in the past and are still continuing.

iv) The plant migration is brought about by the transport of induvidual plants during their mobile dispersal phase.

v) Great variation and oscilation in climate especially at higher lattitude during the geographycal history of angiosperms. Ex. Global warming – Extinction of organisms.

vi) Considerable variation has occurred in the relative distribution and outline (Border) of land and sea in the history of angiosperm and it exerts a high degree of control over distribution of flora.

Theories governing crop adoptation and distribution

1. **Theory of tolerance:** Each and every plant species is able to exist and reproduce successfully only with a definite range of climatic and edaphic conditions. This range represents the tolerance of the species to external conditions.

Shelford (1913) proposed a general law of tolerence which include the following concepts:

- Organisms with wide range of tolerence for all the factors of the environment are likely to be widely distributed.
- Organism may have wide range of tolerence for one factor and a narrow range for another.
- When conditions are not optimum for one factor, the limits of tolerence may be reduced with respect to another factor.
- The period of reproduction is usually critical when environmental factors are likely to be limiting.

Temperature is one of the most limiting factor in plant distribution. Many sub-tropical crops such as wheat, rice willn't withstand high temperature. Among them, rice has wide tolerance range.

2. **Law of minimum (Leibig, 1840):** Refer history
3. **Theory of optima and limiting factor (Blackmann, 1905):** Refer history
4. **Law of Relativity (Lundegardh, 1931):** As a factor increases in intensity, its relative effect on plant growth decreases. This law is considered in connection with the correlation of environmental factors and crop yields.

Origin of Crop Plants: There is a belief that, cultivated crop plants are gifted by god. Also, another belief indicating the process of cultivation itself improved the heredity of plants from the wild.

In 1789, Linnaeus was the first scientist to made effort but, didn't had sufficient information to base a theory.

Decandole (1882), 'Father of Concepts of Cultivated Plants' attempted to make use of earlier information taking evidense from Archeology, Palentology, History and Philosophy. He studied many tree species and tried to trace them back to their wild ancestors through what he called.

Botanical method. The criteria used are:

a) Occurrence of a given cultivated plant in a locality where it is also grown wild or where wild relatives were found.

b) Added information from Archeology, Historical and Linguistic sources.

He studied 247 species including 199 from old world, 45 species from New world and uncertain on 3 species. He published a book "**Origin of Cultivated Plants**".

Nikolai Vavilov (1951), took up extensive survey to locate the possible origin of crop plants and he broughtout a principle that the distribution of plant species on the earth is not uniform attributing to the variation in the geographical factors. He developed differential **Phytogeographical method** to locate the centres of origin of crop plants and listed 8 centres of origin of cultivated plants with the following evidenses:

a) Concentration and great diversity of heritable forms

b) Endemic varietal characters and

c) Presence of closely related wild or cultivated forms.

World centres of origin of cultivated plants

1. **Chinese centre:** Largest independent centre includes the mountainous regions of Central, Western China and adjucent low lands. 136 Plant species were originated here. Ex. Soybean, Pea, Sugarcane (*Saccharum sinensis* L.), Radish, Onion, Cucumber, Peach, Apricot, Millets, Barley, Buck wheat.

2. **Indian centre:** It ha two sub centres.

   a) **Main centre:** Includes Assam and Burma. 117 Plants were originated here. Ex. Rice, Bengalgram, Blackgram,

Greengram, Cowpea, Sugarcane (*Saccharum officinarum* L.), Sesamum, Safflower, Mango, Orrange, Tamarind, Coconut, Brinjal, Jute, Cotton.

b) **Indo-Malayan centre:** Includes Indochina and the Malayan Region. 55 plants are listed here. Ex. Velvet Bean, Banana, Clove, Nutmeg, Black pepper, Bread fruit.

3. **Central Asiatic centre:** Includes NW-India (punjab, J&K), Afganistan, Uzbekistan & Weatern Tiam-Shan. 43 Plants were listed here. Ex. Wheat, Horsegram, Mustard, Garlic, Sunhemp, Jute, Spinach, Carrot, Peas, Grapes, Apple, Almond, Lentil.

4. **Near Eastern centre:** Includes interiro of Asia minor, All of Transcaulasia (Europe), Iran and high lands of Turk Menstan. 83 species are located in this region. Ex. Oat, Barley, Rye, Lucern, Pomogranate, Lupins, Clover, Fenugreek.

5. **Medeterranean centre:** Includes the border of Medetarra-nean Sea. 84 Plants are listed. Ex. Wheat (Durum & Emmer), Flax, Rapeseed, Beet root, Cabbage, Turnip, Celery, Lettuce.

6. **Abyssinian centre:** includes Abyssinia, Eritrea and parts of Somali. 38 Plants are listed. Ex. Sorghum, Bajra, Sesamum, Castor, Ladies finger, Coffee & Millets.

7. **South Mexican and central American Centres:** includes Southern sections of Mexico. Guetamala, Honduras and Costarica. 49 Species are listed here. Ex. Maize, Cotton, Sweet Potato, Guava, Cashew, Papaya, Limabean.

8. **South American centre:** 62 Plants are listed here.

   a) **Peruvian, Equidorean, Bolivian centre:** Comprises mainly the high mountainous areas. Ex. Potato, Tomato, pumpkin, Tobacco, Egyptian cotton.

   b) **Chillian centre:** Island near coast of Southern Chilli. Ex. Wild Straw berry, Potato.

   c) **Brazilian, Paraguayan centre:** Ex. Groundnut, Rubber, Pineapple.

Chapter 15

Crop Management Techniques in Problematic Soils

Problem soils are an important ecological entity of arid, semiarid or humid climate of our country. In India saline and sodic soils are serious threat to produce enough food grains for growing population. The other main problem is acid and acid sulphate soil, spread in one third of cultivated land.

Problem soils are either deficient in plant nutrients or the nutrient availability is inhibited by the antagonistic effect of the nutrient element present in excessive amounts as soluble or exchangeable ions.

1. Soil acidity

1. Acid soils contain excessive amounts of dissolved ions of aluminium, iron, manages and copper.
2. Acid soils are deifict in calcium and magnesium.
3. Acid sulphate soils are rich in pyrites which are oxidation form of sulphuric acid.

Causes for soil acidity

a) Leaching of bases like calcium and magnesium under high rainfall areas reduces soil pH.

b) Continous application of acid forming fertilizer like Muriate of potash (MoP)

$KCl + H_2O \rightarrow KOH + HCl$

c) Decomposition of organic matter liberates several acids.
d) Removal or uptake of bases by the crops.
e) Parent material – some soils might be natively acidic because of associated minerals Ex. Granite.

Management of acidic soils

a) Addition of lime and lime material like oxides, hydroxides and carbonates of Ca & Mg

$$CaCO_3 + H_2O \rightarrow Ca^{2+} + HCO^{3-} + OH^-$$

$$CaO + H_2O \rightarrow Ca^{2+} + HCO^{3-} + OH^-$$

b) Use of basic fertilizers like sodium nitrate and basic slag.
c) Cultivation of acid tolerant crops like sweet potato, castor, rice, barley *etc.*
d) Soil and water management to reduce leaching of bases.

2. Soil Alkalinity:

Alkaline soils are formed in areas which receive low rainfall and accumulation of salts due to less leaching of salts over a period.

Causes for alkalinity

1. High base saturation other than H^+ especially Na, carbonates of Ca & Mg.

$$CaCO_3 + H_2O \rightarrow Ca^{2+} + HCO^{3-} + OH^-$$

2. Presence of excess salts.

Alkaline soils grouped into 3 types based on EC (Electrical conductivity), ESP (Exchangeable Sodium Percentage) & pH values as below:

Class	EC (ds m^{-1})	ESP (%)	pH
1. Saline	>4	<15	<8.5
2. Saline-alkaline / saline sodic	>4	>15	>8.5
3. Alkaline / sodic	<4	>15	>8.5

Electrical conductivity is the measure of dissolved salts in the soils that conduct current approximately proportion to the amount of salt present in the soil.

Reclamation / management

Saline soils

1. Leaching of excessive salts with rain / irrigation.
2. Scraping and removal of accumulated salts in the surface layer.
3. Drainage.
4. Use of acid forming fertiliges like urea, MoP, *etc.*
5. Application of organic manures.
6. Sowing of seeds in the furrows.

Alkaline soils

1. Application of gypsum ($CaSO_4 . 2H_2O$).
2. Addition of Sulphur containing amendments like iron pyrites - FcS_2.

3. Water logged / Marshy lands

In waterlogged areas, stagnation of water occurs for most part of the year. In coastal regions water logging may be caused due to intrusion of sea water especially in east coast. Water logging results in anaerobic conditions beside salt accumulation.

Management options are drainage and cultivation of hydrophytes like rice, trees and grasses.

Major soils of India

1. **Alluvial Soils:** These soils are derived from the deposition laid by the numerous tributaries of the Indus, the Ganges and the Brahmaputra systems. This is the largest and most important soil group of India. Alluvial soils are fertile and suitable for most of the agricultural crops like lowland rice, pulses, cotton, banana *etc.*

2. **Black Soils:** Are dark-grey in colour which is due to the presence of clay-humus complex. Characterized by high swelling and shrinkage property, more plasticity and stickiness, poor drainage and low permeability. Mostly found in Gujarat, Maharashtra, Madya Pradesh and Tamilnadu.

3. **Red Soils:** The red colour of soils is due to the coating of ferric oxides on soil particles. Red soils are light textured, friable, well drained with moderate permeability, Excess gravelliness, surface crusting, and susceptibility to erosion. Occur mostly in semi arid tropics. Southern and eastern parts of peninsular India. Rajasthan, Punjab, Bihar and West Bengal. Suitable crops are maize, wheat, millets, groundnut and pigeon pea.

4. **Lateritic Soils:** Enriched with oxides of Iron and aluminum, low in soil pH, Ca and Mg and rich in organic matter. Well drained and porous and light in texture and found in high rainfall areas of Karnataka, Maharashtra, Kerala, Assam, Eastern ghats and western ghats and Madhya Pradesh. Crops suitable are Rice, coffee, tea, rubber and cashew.

5. **Desert soils:** This soil is formed by broken rocks sand dunes and undulating sandy plains are seen. Clay content is very low (< 8%). High soil pH. Mostly found in Thar Desert and Kutch area. Crops are not raised but in some places dates, palm and coconut tree

6. **Tarai Soils:** Derived from the materials washed down by the erosion of mountains (alluvial origin) hard clay, coarse sand and gravel. High organic matter and high moisture content for greater parts of the year. Sandy loam to silty loam texture, Suitable crops are tall grasses. This type soil found in found in Jammu & Kashmir, UP, Uttaranchal *etc.*

7. **Saline:** These soils contain excess amounts of neutral soluble salts dominating by chlorides and sulphates of Na, Ca and Mg which affect plant growth. White encrustation of salts occur on the surface of the saline soils hence called as **'white alkali'.**

Crops suitable for growing are sesbania, rice, sugarcane, oats, berseem, lucern, barley.

8. **Sodic / Alkaline soils:** These soils contain high amount carbonate and bi carbonates of sodium and these are non saline and with dark encrustation hence called as '**black alkali**'. Soil amelioration with gypsum, green manuring, organic residues, organic manures *etc.*, should be used. Crops suitable Rhodes grass, paragrass, rice, sugar beet. Formed because of poor drainage in arid and semi-arid regions.

9. **Acid soils:** Formed by leaching of bases in high rainfall areas. These soils are low in pH, and high Fe, Al and Mn. Low cat ion exchange capacity high base saturation. Liming and judicious use of fertilizers are the management practices.

10. **Peaty / marshy soils:** Formed in depressions under submerged condition. They have light colour because of ferrous ions and found in Kerala, Parts of Bihar and UP.

Chapter 16

Harvesting and Threshing of Crops

Harvesting is the act of removing crop from where it was growing and moving it to a more secure location for processing and storage. Some root crops and fruits trees can be left in the field or orchard and harvested as needed, but most crops reach a period of maximum quality that is they ripen or mature and will deteriorate if left exposed to the elements. While the major factor determining the time of harvest is the maturity of the crops and other factors such as weather, availability of harvest equipments, packing, drying, transport and storage facilities.

Economic and marketing issues are often even more important to consider when to harvest commodity. Before harvesting of crop, grower must be sure about demand of the produce.

Harvesting the crop at optimum maturity is critical for reaping the benefits of season long efforts on crop production. After final decision on crop maturity, care must be taken to avoid losses during harvest, threshing, transport and storage. Processing is equally important to keep the produce in good condition for realizing good price.

Crop maturity: There are different types of maturity namely physiological maturity, harvest maturity and storage maturity.

Physiological maturity: It is the stage of development in the life cycle of plant when the plant reaches maximum dry weight. At this stage any further increase in inputs does not produce any gain in yield. There is cessation in growth and grain filling at this stage. Grain at this stage has 40 per cent moisture and 90 per cent produce is matured and this remaining 10 per cent still immature. So farmers should harvest at stage to get maximum benefit.

Harvest maturity: This is the stage at which produce is at maximum yield (both quantity and quality). The moisture content will be 18-20 per cent and bring moisture 8-12 per cent depending on type of crop.

Storage maturity: When there is no scope for post harvest drying, the crop is harvested at stage where it can be directly stored. For this produce is left in the field after harvest maturity and moisture will be less than 8-10 per cent depending on different crops.

Methods of harvesting

Different methods of harvesting are followed in different countries, largely depending on timely labour availability and costs associated with harvesting. While mechanical harvesting dominates in developing countries.

Manual harvesting: Manual harvesting is still the major method in developing countries and in India. Certain crops can be harvested without tools, groundnut crop can be harvested by uprooting with hand, provided soil moisture is optimum for hand pulling. Similarly in case with green gram, black gram, horse gram and cowpea.

Mechanical harvesting: Different types of reapers are available like tractor side mounted, tractor front mounted, power tiller operated, self propelled walking and self propelled riding types have been developed for harvesting of crops like soybean, wheat, rice and mustard. Combined harvesting of rice, wheat and soybean has been accepted by farmers in regions with labour shortage during harvesting period.

Threshing and shelling

Threshing or shelling consists of separating the grains or the shells in the case of groundnuts from the portion of the plant that holds them. The separation is done by hand or machine or shaking produce or passing rollers or tractors and hitting against hard surface.

Drying: Drying is the post harvest process where the produce is either shade or sun or artificial dry to bring the moisture per cent to safe storable level. For cereals for moisture per cent should be 10-11 per cent, pulses 9-10 per cent and oil seeds less then 8 per cent.

17

Weeds Importance and Classification

Origin of weeds : Weeds originated together with crop plants. They have been there since the man started cultivating crops about 10,000 BC. Man during his stage of development, explored certain crop plant to suit his own taste and fancy and cultivated them as '**crops**'. The unexplored / unwanted / troublesome plants which are interfering with his activities are considered as weeds. About 30,000 plants species have been identified as definite weeds in the World. Which are infesting crop lands, water bodies, gardens, orchards, road sides, railway lines, irrigation channels *etc.*

Plants are differentiated into

a) **Crops:** Which meet the needs of human being.

b) **Weeds:** Which compete or interfere in human affairs.

Jethrotull is the first person to use the word **weed.**

Several definitions were given on weed and the most common ones are

A plant growing where it is not required or plant out of place or weed is an unwanted plant or weeds are the plants growing in place and at times when we wanted either some other plants to grow or no plants to grow at all.

Weeds are unwanted and undesirable plants that interfere with utilization of land and water resources and thus adversely affect crop production and human welfare. Weeds compete with crop plants for water, nutrients, light, space / atmosphere (CO_2) and thus reduce the crop yields.

A plant of oat in the field of wheat can also be a weed. However, a plant which doesn't interfere with the utilization of land, water and other resources can't be considered as a weed. Hence, plants listed as weeds are qualified based on the situation.

Weediness is a situation in which there is abundance of weeds. Weeding is farm operation where manual and mechanical methods of weed management are followed to provide best opportunity for the crops to establish and grow vigorously up to the harvest.

Importance of weeds

1. **As manure:** When weeds are ploughed, they add plenty of humus to the soil. Excellent compost can be made out of many weed plants. Ex. *Calotropis gigantea*, *Croton sparsiflorus* and *Tephrosia purpurea* are used as green leaf manure for rice.

2. **As human food:** Weeds serve as human food Ex. *Amaranthus viridis* and *Digera arvensis* used as greens.

3. **As fodder:** Most weeds are eaten by cattle and weeds like *Rynchosia aurea*, *R. capitata* and *Clitoria terneata* are very good fodder legumes.

4. **As fuel:** *Prosopis juliflora* very invasive in nature and notorious tree weed commonly used as fire wood. People make charcoal out of it and are marketed.

5. **As soil binders:** *Panicum repens* is an excellent soil binder; keeps bunds in position and prevents soil erosion. We can also use Hariyali (*Cynodon doctylon* L.).

6. **As medicine:** Many weeds have great therapeutic properties and used as medicine as given below:

   a) *Phyllanthus niruri* - Jaundice

   b) *Leucas aspera* - Snake bite

   c) *Centella asiatica* - Improves memory

d) *Eclipta alba* - Scorpion sting

e) *Cynodon dactylon* - Asthma, piles

f) *Cyperus rotundus* - Stimulates milk secretion

7. **As mats and screens:** Stems of *Cyperus spp* are used for mat making while *Typha angustata* is used for making screens.

8. **As indicators:** Weeds are useful indicators of soil fertility Ex. *Echiniclova colonum* occurs in fertile rich soils while *Cymbopogon* occur in poor soil and sedges are found in ill-drained soils.

9. **As pesticide use:** Incorporation of *Crotalaria, Parthenium, Calotrophis, Eichornea spp* into the soil reduces root knot nematode population.

10. **Some of the weeds have economic importance:** Kans (*Saccharum spontaneum* L.) used for thatching purpose and breeding sugarcane varieties for including hardiness.

    Nutgrass / Nutsedge For making essence sticks (Udbatties).

11. **Reclamation of alkali soils:** The application of powder of the weed stayanashi piwala dhotra (*Argemone Mexicana* L.) @ 2.5 tonnes/ha is useful for reclamation of alkali soils.

12. **Some of the weeds serves as ornamental and hedge plants:** Ghaneri (*Lantana camara* L.) and Cactus used as ornamental and hedge plants. Ghol (*Portulaca* spp) for beautiful flowers.

13. **Weeds** can be used for paper pulp, Bio-gas and manufacture of edible proteins.

14. **Some of the weeds** are used for religious purpose Ex. Hariyali, Aghada, Maka *etc.*

It is widely known that, losses caused by weeds exceed the losses from any other agricultural pests such as insects, diseases. In India, weeds accounts for a losses as high as 45 per cent, Insects-30 per cent, Diseases-20 per cent and others-5 per cent. Reduction in yield due to weeds is highest in sugarbeet (75 %) followed by onion (68 %), rice (42 %), maize (40 %), sugarcane (34 %) and groundnut (33 %).

Losses caused by weeds

1. **Reduction in crop yield:** Weeds compete with crop plants for nutrients, soil moisture, space and sunlight. In general an increase in one kilogram weed growth corresponds to reduction in one kilogram of crop growth. The loss in crop yield varies depending on type of weed, intensity of infestation, period of infestation, the ability of crop to compete and climatic conditions.

2. **Reduction in quality of farm produce:** A crop may have to be rejected if it contains weed seeds when it is grown for seed purpose. For example, the wild oat weed seeds are similar in size and shape of the crops like barley, wheat and its admixture may lead to rejection for seed purpose. Contamination by poisonous weed seeds is unacceptable and increases costs of crop cleaning and fetches low price. The leafy vegetables much suffers due to weed problem as the leafy weed mixture spoil the economic value. In tea plantation, presence of Loranthus (*Dendrophthoe falcate* L.) leaves impairs its quality. In cotton, dry weed fragments adhere to its lint and hinder its ginning process.

3. **Weeds harbour pest and diseases and in turn increase cost of production:** Weeds serve as alternate hosts to several insect pests, nematodes and pathogens. Insect such as aphids, thrips, weevils *etc.*, survive on weeds during off-season and transmit to crops in regular season and thus increase the cost of their management. Some example of weeds acting as alternate host of crop pest and diseases are:

Crop	Pest	Alternate Host
Red gram	Gram caterpillar	Amaranthus, Datura
Castor	Hairy caterpillar	Crotalaria sp.
Rice	Stem Borer	Echinocloa, Panicum
Wheat	Black Rust	Agropyron repens
Pearn millet	Ergot	Cenchrus ciliaris
Maize	Downy mildew	Sacharum spontaneum

4. **Interference in crop handling:** Some weeds can make the operation of agricultural machinery more difficult, more costly and even impossible. Heavy infestation of *Cynadon dactylon* causes poor ploughing performance.

5. **Reduction in land value:** Heavy infestation by perennial weeds could make the land unsuitable are less suitable for cultivation resulting in loss in its monetary value. Thousands of hectare of cultivable area in rice growing regions of India have been abandoned or not being regularly cultivated due to severe infestation of nutgrass (*Cyperus rotundus* L.) and other perennial grasses.

6. **Limitation of crop choice:** When certain weeds are heavily infested, it will limit the growth of a particular crop. The high infestation of parasitic weeds such as *Striga spp* may limit the growing of sorghum or sugarcane. While, *Orabanche* limits the choice of tobacco, tomato and other solanaceous crops.

7. **Loss of human efficiency:** Weeds reduce human efficiency through physical discomfort caused by allergies and poisoning. Weeds such as congress weed (*Parthenium hysterophorus* L.) causes itching. Thorny weeds like *Amaranthus spinosus, Argemone mexicana etc.*, restrict the moment of farm workers in carrying out farm practices such as fertilizer application, insect and disease control measures, irrigation, harvesting *etc*.

8. **Problems due to aquatic weeds:** The aquatic weeds that grow along the irrigation canals, channels and streams restrict the flow of water. Weed obstruction cause reduction in velocity

of flow and increases stagnation of water and may lead to high siltation, flooding, seepage to adjoining areas, break canal banks, and reduced carrying capacity. Aquatic weeds form breeding grounds for obnoxious insects like mosquitoes.

9. **Deteriorate the aesthetic value:** They reduce recreational value by interfering with fishing, swimming, boating, hunting and navigation on streams and canals. Presence of weeds around living and working places makes the surrounding dull and insipid.

10. **Weed menace in animal husbandry:** When some hungry animals are forced to feed on weedy forage, their meat and milk get odd flavoured and tainted. Certain weeds like *Euphorbia simplex, Lepidium sativa, Cleome viscose* impart an undesirable smell to milk upon feeding to cattle. Certain weeds like *Cenchurus spp., Achyranrthus aspera* entangled to hairs/wool of animals (sheep) and reduce the quality of wool.

    Certain weeds cause sickness in farm animals due to the presence of alkaloids, tannins, oxalates, glucosides or nitrates. Johnson grass (*Sorghum halopense* L.) at its tillering stage and *Xanthium pungens* at its cotyledonous stage is poisonous to animal due to their high Prussic acid content.

11. **Weed menace in human health:** Health, comfort and work efficiency of man are adversely affected by weeds. Numerous people are plagued year after year with hay fever and asthma aggravated by pollens of *Ambrosia artimissifolia* and *Fraseria spp. Parthenium hysterophorus* is responsible for allergy. Mexican poppy *(Argemone mexicana)* seeds crushed with mustard seeds have brought death and blindness to thousands of people.

12. **Weed menace to industry and public utilities:** Weeds growing in industrial sites are potential source of fire hazards. Also, they block pipelines, roads line, electric poles *etc.,* reduce their easy access.

13. **Other problems:** Weeds are troublesome not only in crop plants but also in play grounds and road / railway sides *etc. Alternanthera echinata* and *Tribulus terresstris* occurs in many of the playgrounds causing annoyance to players and spectators.

Classification of weeds: There are many ways weeds can be grouped for convenience of planning, interpreting or suggesting appropriate weed management practices. Some important classifications of weeds used by weed scientists worldwide are discussed below:

I. Based on life span or Ontogeny

Based on life span (Ontogeny), weeds are classified as Annual weeds, Biennial weeds and Perennial weeds:

1. **Annual weeds:** Those that live only for a season or year and complete their life cycle in that season or year is called annuals. These are small herbs with shallow roots and weak stem. Produces seeds in profusion and the mode of propagation is commonly through seeds. After seeding the annuals die away and the seeds germinate and start the next generation in the next season or year following.

Based on the growing season, annuals are further classified as :

i) ***Kharif* annuals:** *Leucas aspera, Digitaria marginata, Eleusine indica.*

ii) **Winter or *rabi* annuals:** *Phalaris minor, Avena fetua, Chenopodium album.*

iii) **Summer annuals:** *Amaranthus spp.* and most of *kharif annuals.*

Plants which complete their life cycle with in very short span are called ***ephemerals.*** Ex. *Phyllanthus niruri* L.

2. **Biennials:** It completes the vegetative growth in the first season, flower and set seeds in the succeeding season and then dies. These are found mainly in non-cropped areas. They are easy to manage as we can control during vegetative period. Ex. *Alternanthera echinata* L., *Daucus carota* L., *Melilotus alba* L.

3. **Perennials:** Perennials live for more than two years and may live almost indefinitely. They adapted to withstand adverse conditions. They propagate not only through seeds but also by underground stem, root, rhizomes, tubers *etc., a*nd hence they are further classified into

   a) **Simple perennials:** Plants propagated only by seeds. Ex. *Sonchus arvensis* L.

   b) **Bulbous perennials:** Plants which possess a modified stem with scales and reproduce mainly from bulbs and seeds. Ex. *Allium* sp.

   c) **Corm perennials:** Plants that possess a modified shoot and fleshy stem and reproduce through corm and seeds. Ex. *Timothy* sp., wild garlic, agave.

   iv) **Creeping perennials:** Reproduced through seeds as well as with one of the following.

   a) **Rhizome:** Plants having underground stem – *Sorghum halapense* L.

   b) **Stolon:** Plants having horizontal creeping stem above the ground – *Cynodon dactylon* L.

   c) **Roots:** Plants having enlarged root system with numerous buds – *Convolvulus arvensis* L.

   d) **Tubers:** Plants having modified rhizomes for storage of food – *Cyperus rotundus* L.

II. Based on ecological origin

1. **Wetland weeds:** They are tender annuals with semi-aquatic habit. They can thrive well under waterlogged as well as in dry condition. Propagation is chiefly by seed. Ex. *Ammania baccifera* L. and *Eclipta alba* L.
2. **Garden land weeds (Irrigated lands):** These weeds neither require large quantities of water like wetland weeds nor can they successfully withstand extreme drought as dryland weeds Ex. *Trianthema portulacastrum* L. *and Digera arvensis* L.
3. **Dry lands weeds:** These are usually hardy plants with deep root system. They are adapted to withstand drought on account of mucilaginous nature of the stem and hairiness. Ex. *Tribulus terrestris* L., *Convolvulus arvensis* L.

III. Based on soil type (Edaphic)

1. **Weeds of black cotton soil:** These are often closely allied to those that grow in dry condition. Ex., *Aristolochia bracteata* L.
2. **Weeds of red soils:** They are like the weeds of garden lands consisting of various classes of plants. Ex. *Commelina benghalensis* L.
3. **Weeds of light, sandy or loamy soils:** Weeds that occur in soils having good drainage. Ex. *Leucas aspera* L.
4. **Weeds of laterite soils:** Ex. *Lantana camara* L., *Spergula arvensis* L.

IV. Based on place of occurrence

1. **Crop land weeds:** The majority of weeds infests the cultivated lands and cause hindrance to the farmers for successful crop production. Ex. *Phalaris minor* L.in wheat.
2. **Plantation weeds:** Occurring in plantation area Ex.: *Lantana camara* L *and, Commelina bengalensis* L.

3. **Forest and wood land weeds**

4. Garden land weeds

5. **Weeds of Grass land or Range land or pasture lands:** Weeds found in pasture / grazing grounds. Ex. *Indigofera enneaphylla* L.

6. **Weeds of fallow and non-cropped waste places:** Corners of fields, margins of channels, industrial areas etc., where weeds grow in profusion. Ex. *Gynandropsis pentaphylla* L. *and Calotropis gigantean* L.

7. **Weeds of playgrounds and road-sides:** They are usually hardy, prostrate perennials, capable of withstanding any amount of trampling. Ex. *Alternanthera echinata* L., *Tribulus terestris* L.

8. **Aquatic weeds:** Ex: *Eicorrnea crossipes* L.

V. Based on Origin

1. **Indigenous weeds:** All the native weeds of the country are coming under this group and most of the weeds are indigenous. Ex. *Acalypha indica* L., *Abutilon indicum* L.

2. **Introduced or Exotic weeds:** These are the weeds introduced from other countries. These weeds are normally troublesome and control becomes difficult. Ex., *Parthenium hysterophorus* L., *Phalaris minor* L., *Acanthospermum hispidum* L.

VI. Based on cotyledon character

Based on number of cotyledons it possess it can be classified as dicots and monocots:

1. **Monocots** Ex. *Panicum flavidum* L., *Echinochloa colona* L.

2. **Dicots** Ex. *Crotalaria verucosa* L., *Indigofera viscose* L.

VII. Based on soil pH

Based on pH of the soil the weeds can be classified into three categories :

a) Acidophile: Acid soil weeds Ex. *Rumex acetosella* L.

b) Basophile: Saline and alkaline soil weeds Ex. *Traxacum stricta* L.

c) Neutrophile: Weeds of neutral soils ex. *Acalypha indica* L.

VIII. Based on morphology

Based on the morphology of the plant, weeds are also classified in to three categories. This is the most widely used classification by the weed scientists :

1. **Grasses:** All the weeds come under the family Poaceae are called as grasses which are characteristically having long narrow spiny leaves. The examples are *Echinocloa colonum* L., *Cynodon dactylon* L.
2. **Sedges:** The weeds belonging to the family Cyperaceae come under this group. Stems are solid and triangular in cross section. The leaves are mostly from the base having modified stem with or without tubers. They are distinguished from grasses by having no joints in the stem. They grow in marshy or swampy areas. The examples are *Cyperus rotundus* L., *Fimbrystylis miliaceae* L.
3. **Broad leaved weeds:** All dicotyledonous weeds are broad leaved with netted venation. The examples are *Flavaria australacica* L. and *Digera arvensis* L.

IX. Based on nature of stem

Based on development of bark tissues on their stems and branches, weeds are classified as woody, semi-woody and herbaceous species:

1. **Woody weeds:** Weeds include shrubs and undershrubs and are collectively called brush weeds. Ex. *Lantana camera and Prosopis juliflora* L.

2. **Semi-woody weeds:** Ex.*Croton sparsiflorus* L. *Cannabis sativa, Acahspha* sp. *Parthemium histero.*

3. **Herbaceous weeds:** Weeds have green, succulent stems are of most common occurrence around us. Ex. *Amaranthus viridis* L. *Chenopodium album*, *Melicotus alba*.

X. Based on specificity

Besides the various classes of weeds, a few others deserve special attention due to their specificity. They are:

1. Poisonous weeds
2. Parasitic weeds
3. Aquatic weeds

1. **Poisonous weeds:** The poisonous weeds cause ailment on livestock resulting in death and cause great loss. These weeds are harvested along with fodder or grass and fed to cattle or while grazing the cattle consumes these poisonous plants. Ex. *Datura fastuosa* L., *D. stramonium* L. and *D. metel* L. are poisonous to animals and human beings. The berries of *Withania somnifera* L. and seeds of *Abrus precatorius* L. are poisonous.

2. **Parasitic weeds:** The parasite weeds are either total or partial which means, the weeds that depend completely on the host plant are termed as total parasites while the weeds that partially depend on host plant for minerals and capable of preparing its food from the green leaves are called as partial parasites.

Those parasites which attack roots are termed as root parasites and those which attack shoot of other plants are called as stem parasites. The typical examples of different parasitic weeds are;

1. **Total root parasite:** *Orabanche cernua* L. on Tobacco.

2. **Partial root parasite:** *Striga lutea* L. on sugarcane and sorghum.

3. **Total stem parasite:** *Cuscuta chinensis* L. on leucerne and onion.

4. **Partial stem parasite:** *Cassytha filiformis* on orange trees and *Loranthus longiflorus* L. on mango and other trees.

3. **Aquatic weeds:** Unwanted plants, which grow in water and complete at least a part of their life cycle in water are called as aquatic weeds. They are further grouped into four categories as submerzed, emerged, marginal and floating weeds.

   - **Submerged weeds:** These weeds are mostly vascular plants that produce all or most of their vegetative growth beneath the water surface, having true roots, stems and leaves. Ex. *Utricularia stellaris* L. *and Ceratophyllum demersum* L.

   - **Emerged weeds:** These plants are rooted in the bottom mud, with aerial stems and leaves at or above the water surface. The leaves are broad in many plants and sometimes like grasses. These leaves do not rise and fall with water level as in the case of floating weeds. Ex. *Nelumbium speciosum* L. and *Jussieua repens* L.

   - **Marginal weeds:** Most of these plants are emerged weeds that can grow in moist shoreline areas with a depth of 60 to 90 cm water. These weeds vary in size, shape and habitat. The important genera that come under this group are; *Typha, Polygonum* and *Cephalanthus,etc.*

   - **Floating weeds:** These weeds have leaves that float on the water surface either singly or in cluster. Some weeds are free floating and some rooted at the mud bottom and the leaves rise and fall as the water level increases or decreases. Ex. *Eichhornia crassipes* L., *Pistia stratiotes* L., *Salvinia, Nymphaea pubescens* L.

XI. Classification according to association

According to association, weeds are grouped into 3 types :

1. **Season bound weeds:** These weeds grow in specific season disregard to the crop species cultivated. These weeds may be

either summer annuals Ex. *Sorghum halepence* L. or winter annuals *Circium arvensis* L.

2. **Crop bound weeds:** Are those species of weeds which are usually parasitize the host crops. They depend upon the host plants for nutrition.

3. **Crop associated weeds:** Non-parasitic weeds associated with specific crop. They may be associated with crop plants for one or more of following reasons.

   i) **Need for specific micro-climate:** Weeds like *Chicorium intybus* L. (Chichory) require shady, cool and moist habitat which is amply available in crops like berseen and Lucerne.

   ii) **Mimicry:** Wild rice in paddy and *Phalaris minor* L. in wheat survive because of their morphological similarity with host plants. So is true for Loranthus in Tea. This mechanism is called mimicry.

   iii) **Ready contamination of crop seed:** *Allium spp,* Wild garlic, *Phalaris minor* L. mature and spread their seeds same height and time of main crop and thus easily contaminate with crop seeds at harvesting time.

XII. Based on association with human affairs

1. **Facultative weeds:** Are those weed species that grow primarily in wild communities but often escape to cultivated fields associating themselves with human affairs. Ex. Wild onion.

2. **Obligate weeds:** Occur only in cultivated field. They can't withstand competition from volunteer vegetation in a closed community (dominant weeds). Ex. *Convolvulus arvensis.*

XIII. Noxious and objectionable weeds

The noxious weeds are very difficult to control because of their high competitive nature. The weeds persist in adverse conditions and reduce the crop yield even at very low density. The following are considered as the 10 worst weeds of the world.

1. *Cyperus rotundus* L. - Purple nut sedge
2. *Cynodon doctylon* L. - Bermuda grass
3. *Echinocloa crusgalli* L. - Jungle rice
4. *Sorghum halepense* L. - Johnson grass
5. *Eichornia crassipes* L. - Water hyacinth
6. *Imperata cylindrica* L. - Quack grass
7. *Lantana camera* L. - Lantana
8. *Parthenium hysteroporus* L. - Parthenium
9. *Echinocloa colanum* L. - Barn yard grass
10. *Convolvulus arvensis* L. - Field bind weed

An objectionable weed is a troublesome weed whose seeds are difficult to separate once mixed with crop seeds.

Crop weed competition and crop weed interference

Struggle for survival and existence is called competition. Water, nutrients, solar radiation and space are the major factors for which competition usually occurs. Competition between crop plants and weeds is most severe when they have similar growth habit and demand for common growth factors.

Dominance of a habitat by the crop or weed depends on the rapidity of seed germination, seedling establishment and subsequent growth and development. Differences in photosynthetic area, root development and its growth largely determine the competitiveness of plants. The degree of weed completion is determined by the weed species infecting the crop, density of infestation and duration of infestation.

1. **Competition for Nutrients:** Weeds usually absorb mineral nutrients faster than many crop plants and accumulate them in their tissues in relatively larger amounts. For instance, *Amaranthus* sp. accumulate over 3 % N on dry weight basis and are termed as "nitrophills". *Acharanthus aspera*, a 'P' accumulator with over 1.5 % P_2O_5, *Chenopodium* sp & *Portulaca* sp. are 'K' lovers with over 1.3 % K_2O in dry matter.

Further, the nutrient content of our dominant crops varies between 0.33-1.33 % N, 0.19-0.59 % P_2O_5 and 0.67-1.44 % K_2O.

2. **Competition for moisture:** In general, for producing equal amounts of dry matter, weeds transpire more water than do most of our crop plants. It becomes increasingly critical with increasing soil moisture stress, as found in arid and semi-arid areas. *Cynodon dactylon* had almost twice as high transpiration rate as pearl millet.

The amount of water required to produce unit quantity of dry matter is known as transpiration co-efficient (T_Q). The weeds have 2-3 time higher T_Q than crop plants.

Table : Transpiration coefficient (T_Q) of crops and weeds

Weed	T_Q	Crop plant	T_Q
Cynodon doctylon	813	*Zea mays* L.	352
Tridox procumbens	1402	*Sorghum bicolor* L.	394

3. **Competition for light:** Generally weeds grow faster and shade the plants if not checked. Even in case of shorter weeds, the lower leaves of crop plants are shaded and deprived of light for photosynthesis. Unlike competition for nutrients and moisture once weeds shade a crop plant, increased light intensity cannot benefit it. The crop plants become thin and pale green.

4. **Competition for CO_2:** Crop-weed competition for CO_2 may occur under extremely crowded plant community condition. A more efficient utilization of CO_2 by C4 type weeds may contribute to their rapid growth over C3 type of crops. Also, weeds can thrive well even under low CO_2 concentration.

Critical period of crop weed competition

Critical period of weed competition is defined as the shortest time span during the crop growth when weeding results in highest economic returns. Crop yield reduction due to weeds is maximum during this stage. To get higher yield weed free environment during this stage is important.

Critical period of competition is the period from sowing up to which the crop has to be maintained in a weed free environment for remunerative crop production.

The crop canopy in the early period of growth is inadequate to smother weed growth. By early reproductive phase, the crop develops leaf area adequate to smother the weed growth. Hence, weed competition in crop field is invariably severe in early stages of crop growth than at later stages. Generally in the crop of 100 days duration, the first 35 days of sowing should be maintained weed free for optimum yield. In general, crops must be maintained weed free during first one third period of life cycle.

Critical period of weed competition for important crops

1. Rice (Lowland)	- 35 days	7. Cotton	- 35 days
2. Rice (upland)	- 60 days	8. Sugarcane	- 90 days
3. Sorghum	- 30 days	9. Groundnut	- 45 days
4. Finger millet	- 35 days	10. Soybean	- 45 days
5. Pearl millet	- 35 days	11. Onion	- 60 days
6. Maize	- 30 days	12. Tomato	- 30 days

It is clear that weed free condition for 2-8 weeks in general are required for different crops and emphasizes the need for timely weed control without which the crop yield gets drastically reduced.

Factors affecting crop-weed competition

a) **Type of weeds species:** The type of weeds that occur in a particular crop influences the competition. Occurrence of a particular species of weed greatly influence the competition between the crop & weed. Ex. *E. crusgalli* in rice, *Setaria viridis* in corn and *Xanthium* sp. in soybean affects the crop yield.

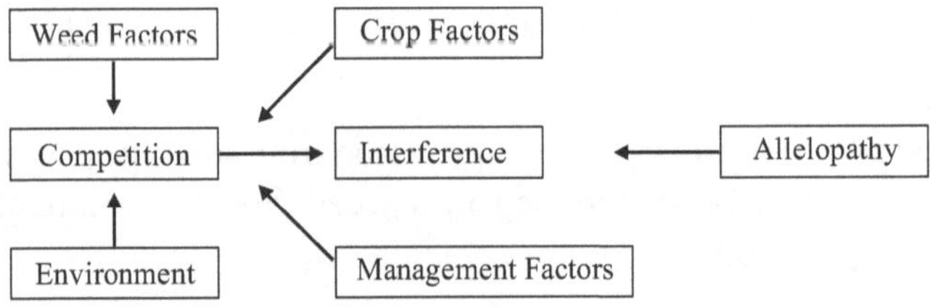

b) **Type of crop species and their varieties:** Crops and their varieties differ in their competing ability with weeds Ex. decreasing order of weed competing ability is as: barley, rye, wheat and oat. High tolerance of barley to competition from weeds is assigned to its ability to develop more roots that are extensive during initial three weeks growth period than the others.

Fast canopy forming, tillering and tall crops suffer less from weed competition than the slow growing, non-tillering and short stature & crops. Dwarf and semi-dwarf varieties of crops are usually more susceptible to competition from weeds than the tall varieties because they grow slowly during initial stage. When we compare the crop-weed competition between two varieties of groundnut *viz.*, Bunch and Spreading, bunch type incurred a loss of over 30 per cent pod yield under uncontrolled weed-crop competition while spreading type lost only about 15 per cent in its yield. The main reason is due to the spreading nature, which smothered weeds. Longer duration cultivars of rice have been found more competitive to weeds than the short duration ones.

c) **Density of weeds:** Increase in density of weed decrease in yield is a normal phenomena. However, it is not linear as few weeds do not affect the yields so much as other weed does and hence, it is a sigmoidal relationship.

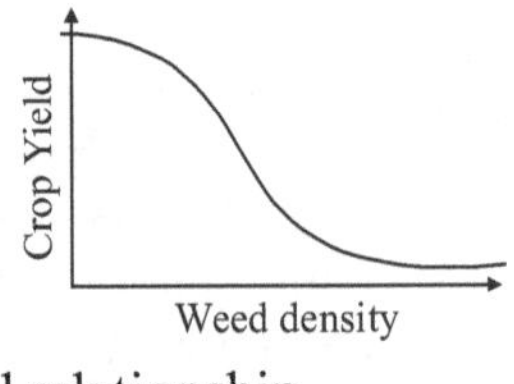

d) **Plant density:** Increase in plant population decreases weed growth and reduces competition until they are self competitive. Crop density and rectangularity are very important in determining the quantum and quality of crop environment available for the growth of weeds. Wide row spacing with simultaneous high, intra-row crop plant population may induce dense weed growth. In this respect, square planting of crops in which there are equal row and plant spacing should be ideal in reducing intra-crop plant competition.

e) **Soil factors:** Soil type, soil fertility, soil moisture and soil reaction influences the crop weed competition. Elevated soil fertility usually stimulates weeds more than the crop, reducing thus crop yields. Fertilizer application of weedy crop could increase crop yields to a much lower level than the yield increase obtained when a weed free crop is applied with fertilizer.

Weeds are adapted to grow well and compete with crops, in both moisture stress and ample moisture conditions. Removal of an intense moisture stress may thus benefit crops more than the weeds leading to increased yields. If the weeds were already present at the time of irrigation, they would grow so luxuriantly as to completely over cover the crops. If the crop is irrigated after it has grown 15 cm or more in a weed free environment irrigation could hasten closing in of crop rows, thus suppressing weeds.

Abnormal soil reactions often aggravate weed competition. It is therefore specific weed species suited to different soil reactions exist with us, our crops grow best only in a specified range of soil pH. Weeds would offer more intense competition to crops on normal pH soils than on normal pH soils.

f) **Climate:** Adverse weather condition, Ex. drought, excessive rains, extremes of temperature, will favour weeds since most of our crop plants are susceptible to climatic stresses. It is further intensified when crop cultivation is stratified over marginal lands. All such stresses weaken crops inherent capacity to fight weeds.

g) **Time of germination:** In general, when the time of germination of crop coincides with the emergence of first flush of weeds, it leads to intense Crop-Weed interference. Sugarcane takes about one month to complete its germination phase while weeds require very less time to complete its germination. Weed seeds germinate most readily from 1.25 cm of soil. Few weeds even from 15 cm depth. Therefore, planting method that dries the top 3 to 5 cm of soil rapidly enough to deny weed seeds opportunity to absorb moisture for their germination usually postpones weed emergence until the first irrigation. By this time the crop plants are well established to compete with late germinating weeds.

h) **Cropping practices:** Cropping practices, such as method of planting crops, crop density and geometry and crop species and varieties have pronounced effects on Crop-Weed interference.

i) **Crop maturity:** Maturity of the crop is yet another factor which affects competition between weeds & crop. As the age of the crop increases the competition for weeds decreases due to its good establishment. Timely weeding in the early growth stages of the crop enhances the yield significantly.

Crop Weed Interference

The weed competetion for nutrient, water, light and space *etc*., is considered as an indirect effect on the crop growth & yield. On the other hand the weeds also release whole chemicals & affet the crop growth & yield by an phenomenon called as allelopathy. It is a direct effect on crop growth & yield.

Crop Weed Competition however does not include allelopathy which is also a possible mechanism for reduction in growth & yield of crop in field. In edition the detremental effect of one plant species on another resulting from interaction with each other is called as interference. Therefore the total detremental effect of weeds on crop plants resulting in interacting with each other is termed as crop weed Interference (Harpar, 1916). Which includes bothcompetition & Allelopathy. Therefore, Crop Weed Interference is more scientific and appropriate than crop weed competition.

Concept of weed management-principles and methods

Concept of weed management deals with decision making process in weed management while the principles of weed management aim at points to be considered for effective weed management.

Concept in weed management

There are only three concepts that apply in weed management :

1. Control at any cost
2. Control at economic threshold level
3. Control at ecological threshold level

Control at any cost: The concept of control at any cost aims to maintain level of weed infestation as low as possible in order to prevent yield reduction and avoid further increase in weed infestation. This is the oldest concept of weed control and has been probably practiced since the early days of agriculture when truly no effective weed control methods were available and a large portion of farming activity was spent on weeding. This method includes all the possible methods to control weeds including crop rotation, selection of weed free seeds, *etc.,* and its costly and time consuming.

Control at economic threshold level: Control at economic threshold be defined as the point at which economic damage to a crop is about equal to the cost of controlling the weeds. In this weeds are tolerated to a certain extent and the farmer need not be afraid of being unable to manage an increased infestation in the coming year because he has not controlled his weeds in the previous season.

Control at ecological threshold level: The concept of ecological thresholds is much more complicated and goes beyond the economically based question. How many weed can we tolerate and how many weeds we need. At the moment we know far too little to use this concept. Weeds also useful and beneficial to human beings.

Principles of weed management

For successful weed control, one has to consider the following principles:

1. **Habitat of weed plants:** Xerophytic weed thriving under dry and arid conditions will die if fields are flooded with water. Similarly, weeds which thrive under marsh or ill drained soil condition can be controlled by improving drainage.
2. **Life cycle of weeds:** Annual and biennials can be controlled effectively if the land is cultivated before seedling stage of weeds. Perennials require deep ploughing to dig out vegetative parts.
3. **Susceptibilities:** Some weeds are susceptible to certain chemicals while others are not. Dicots are susceptible to 2.4-D, while monocots are not and hence 2.4-D is used to control broad leaved weeds in monocots crops.
4. **Dormancy period:** While controlling weeds, dormant period is to be considered.
5. **Resistant to adverse conditions without losing viability:** Some weed seeds have hard seed coat which enables them to remain dormant for a long time without losing their viability. Such weeds should be controlled before seed formation.
6. **Method of reproduction:** Propagation either by seed or vegetative parts or by both. Propagation through seed weed should be removed before seed formation stage. Vegetative propagated weed should be controlled by exposing rhizomes, bulbs, stoles, *etc*., by deep plough.

Methods of weed management

1. Prevention
2. Eradication
3. Control / management

Prevention: It encompasses all measures taken to prevent the introduction, establishment and spread of weeds in local or regional or national level. No weed control programme is successful if adequate preventive measures are not taken to reduce weed infestation. Following preventive control measures are suggested for adoption wherever possible and practicable.

1. Clean cultivation.
2. Use of weed free clean and certified seeds.
3. Keeping seed beds free from weeds.
4. Avoid feeding material containing weed seeds to the farm animals.
5. Avoid adding weeds which set seeds to the manure pits.
6. Clean the farm machinery thoroughly before use.
7. Keep irrigation channels, fence-lines, and un-cropped areas clean.
8. Use vigilance. Inspect the farm frequently for any strange looking weed seedlings. Destroy such patches of a new weed by digging deep and burning the weed along with its roots.
9. Quarantine regulations deny the entry of weed seeds and other propagules from one country to another at airports and ship yards.
10. Remove / destroy weeds before they attain reproductive stage or setting seeds.
11. Prevention by weed Laws.

Eradication: It refers that a given weed species, its seeds and vegetative part have been killed or completely removed from a given area and that weed will not reappear unless reintroduced to the area. Because of its difficulty and high cost, eradication is usually attempted only in smaller areas such as few hectares or a few thousand square meters or less. Eradication is often used in high value areas such as green houses, ornamental plant beds and containers. This may be desirable and economical when the weed species is extremely noxious and persistent as to make cropping difficult and economical.

Weed Control / Management: It encompasses those processes where by weed infestations are reduced but not necessarily eliminated. Weed control aims at only putting down the weeds present by some kind of physical or chemical means while weed management is a system approach where in advance planning is done to minimize the invasion of weeds in aggressive forms and give crop plants a strongly competitive advantage over the weeds.

Weed control / management methods are grouped into cultural (agronomical), physical, chemical and biological. Every method of weed control has its own advantages and disadvantages. No single method is successful under all weed situations. Many a time, a combination of these methods gives effective and economic control than a single method.

Physical or Mechanical weed control / management

Mechanical or physical methods of weed control are being employed ever since man began to grow crops. The mechanical methods include tillage, hoeing, hand weeding, digging cheeling, sickling, mowing, burning, flooding, mulching *etc.*

1. **Tillage:** Tillage removes weeds from the soil resulting in their death. It may weaken plants through injury of root and stem pruning, reducing their competitiveness or regenerative capacity.Tillage also buries weeds and also exposes weed seeds to the soil surface for solar scorching. Tillage operation includes ploughing, discing, harrowing and leveling. In case of perennials, both top and underground growth is injured and destroyed by tillage.

Tillage operations are broadly grouped as pre plant and post plant / inter-cultural operations which are aiming for different purposes *viz.*, soil moisture conservation, aeration, incorporating the manure and indirectly promote the crop growth for subsequent suppression of weeds.

2. **Summer tillage:** The practice of summer tillage or off-season tillage is one of the effective cultural methods to check the growth of perennial weed population in crop cultivation. Initial tillage before cropping should encourage clod formation. These clods, which have the weed propagules, upon drying desiccate the same. Subsequent tillage operations should break the clods into small units to further expose the shriveled weeds to the hot sun.

3. **Hoeing:** Hoe has been the most appropriate and widely used weeding tool for centuries. It is however, still a very useful implement to obtain results effectively and cheaply. Hoeing is particularly more effective on annuals and biennials as weed growth can be completely destroyed. In case of perennials, it destroy the top growth with little effect on underground plant parts resulting in re-growth.

4. **Hand weeding:** It is done by physical removal or pulling out of weeds by hand or removal by implements called khurpi. It is probably the oldest method of controlling weeds and it is still a practical and efficient method of eliminating weeds in cropped and non-cropped lands. It is very effective against annuals, biennials and controls only upper portions of perennials. The major set back here is the labour availability, time consumption and ideal soil condition.

5. **Digging:** Digging is very useful in the case of perennial weeds to remove the underground propagating parts of weeds from the deeper layer of the soil. This can be done using pickaxe or guddali or crow bar.

6. **Cheeling:** It is done by hand using a cheel hoe, similar to a spade with a long handle. It cuts and shapes the above ground weed growth. It is useful on annuals and biennials and widely used in plantation crops.

7. **Sickling:** Sickling is also done by hand with the help of sickle to remove the top growth of weeds to prevent seed production and to starve the underground parts. It is popular in sloppy areas where only the tall weed growth is sickled leaving the root system to hold the soil in place to prevent soil erosion.
8. **Mowing:** It is the cutting of weeds at uniform height and used to restrict seed production and check excess weed growth. It is a machine-operated practice mostly done on roadsides and lawns.
9. **Burning or flaming:** Burning or flaming is often an economical and practical means of controlling weeds. It is used to (a) dispose of vegetation (b) destroy dry tops of weeds that have matured (c) kill green weed growth in situations where cultivations and other common methods are impracticable.

Flaming is the exposure of green weeds to a very high temperature (up to 100°C) using flames from burning petroleum gas. Flames are directed towards weeds in between crop rows with a hood cover.

10. **Flooding:** Flooding is successful against weed species sensitive to submergence in water. Flooding kills plants by reducing oxygen availability for plant growth. The success of flooding depends upon level and period of submergence of weeds.
11. **Dredging and chaining:** These methods are adopted for aquatic weeds. Removing of weeds along with their roots and rhizomes with the help of mechanical force is called dredging. The floating aquatic weeds are removed by chaining. High strength chain is pulled over the water bodies to collect weeds.

Merits of physical or mechanical method

1. Oldest, effective and economical method
2. Large area can be covered in shorter time
3. Safe method for environment
4. Does not involve any skill

5. Weeding is possible in between plants
6. Deep rooted weeds can be controlled effectively

Demerits of mechanical method

1. Labour consuming
2. Possibility of damaging crop
3. Requires ideal and optimum soil condition

Agronomical / cultural methods: Several cultural practices like field preporatics, planting, fertilizer application, irrigation selection of variety, time of sowing, cropping system, cleanliness of the farm *etc.*, are employed for creating favourable condition for the crop. These practices are used properly help in controlling weeds. Cultural methods, alone cannot control weeds, but help in reducing weed population. They should, therefore, be used in combination with other methods. In cultural methods, tillage, fertilizer application and irrigation are important apart from :

1. **Field preparation:** The field has to be kept clean and weed free. Flowering of weeds should not be allowed and this helps in prevention of build up of weed seed population. Field operations must be planned in such a way that field is kept weed free.
2. **Varieties:** Short stature and erect leaved varieties permit higher light compared to tall and spreading type of varieties. Weeds continue to germinate for longer time in dwarf varieties resulting to higher infestation.
3. **Sowing / planting methods:** Sowing / planting should be taken with in three days after rainfall or irrigation. Weeds present in soil start germinating with in 2-3 days and suppress the crop plants if we delay the sowing.

Sowing with seed drill followed by passing blade harrow in between the rows to cover seeds will disturb weeds which are already germinated. Transplanting is another operation which reduces the weed growth as the crop has an advantage of age.

4. **Time of planting:** Optimum time of planting crops seems also best season for emergence of weeds. Attributing to the availability of photo-insensitive varieties, in field with a bad history of weeds, crops can be sown / planted either little earlier or later than the normal planting. This helps to avoid first flush of weeds.

5. **Optimum plant population:** Lack of adequate plant population is prone to heavy weed infestation, which becomes difficult to control later. Therefore practices like selection of proper seed, right method of sowing, adequate seed rate *etc.,* very important to obtain proper and uniform crop stand which are capable to offer competition to the weeds.

6. **Crop rotation:** The possibility of a certain weed species or group of species occurring is greater if the same crop is grown year after year. In many instances, crop rotation can eliminate or atleast reduce difficult weed problems. The obnoxious weeds like *Cyperus rotundus* can be controlled effectively by including low land rice in crop rotation. Cotton - sorghum crop rotation to control striga.

7. **Smother or intercrops:** Inter cropping suppresses weeds better than sole cropping and thus provides an opportunity to utilize crops themselves as tools of weed management. Many short duration pulses *viz.*, green gram and soybean effectively smother weeds without causing reduction in the yield of main crop. Smother crop germinates quickly and develop large canopy with deep roots. They suppress weeds by excluding light and utilizing large portion of nutrients, water in the soil.

8. **Mulching:** Mulch is a protective covering of material maintained on soil surface. Mulching has smothering effect on weed control by excluding light from the photosynthetic portions of a plant and thus inhibiting the top growth. It is very effective against annual weeds and some perennial weeds like *Cynodon dactylon.* Mulching is done by dry or green crop residues, plastic sheet or polythene film.

9. **Solarization:** This is another method of utilization of solar energy for the desiccation of weeds. In this method, the soil temperature is further raised by 10-12°C by covering a presoaked fallow field with thin transparent polyethylene sheet during hot summer months. The polyethylene sheet checks the long wave back radiation from the soil and prevents loss of energy by hindering moisture evaporation. Solarization kills the weeds through desiccating the weed seeds, suffocation through ceasing gas exchange, scorching of the sprouted seeds and microbial disintegration / desiccation of weeds seeds.

10. **Stale seedbed:** A stale seedbed is one where initial one or two flushes of weeds are destroyed before planting of a crop. This is achieved by soaking a well prepared field with either irrigation or rain and allowing the weeds to germinate. At this stage a shallow tillage or non- residual herbicide like paraquat may be used to destroy the dense flush of young weed seedlings. This may be followed immediately by sowing. This technique allows the crop to germinate in almost weed-free environment.

11. **Fertilizer and water management:** Plants differ in their ability of responding to applied fertilizer and water. Application of fertilizer and irrigation water at appropriate time, method and depth will improve the crop vigour through their efficient use and suppress the weeds.

Merits of cultural method

1. Low cost for weed control
2. Easy to adopt
3. No residual problem
4. No damage to crops

Demerits of cultural method

1. Immediate and quick weed control is not possible
2. Weeds are kept under suppressed condition
3. Perennial and problematic weeds cannot be controlled.

Biological weed control / management: It is the control or suppression of weeds by the action of one or more organisms through natural means or by manipulation of weed, organism or environment.

Use of living organism's *viz.*, insects, and disease causing organisms, herbivorous fish, snails or even competitive plants for the control of weeds is called biological control. In biological control method, it is not possible to eradicate weeds but weed population can be reduced. This method is not useful to control all types of weeds. Introduced weeds are best targets for biological control.

Qualities of bio-agent

1. **Host specific:** The bio-agent must feed or affect only on the targeted host.
2. It must be free of predators or parasites.
3. It must adapt readily to the environment conditions.
4. Must be capable of seeking out itself to the host.
5. It must kill the targeted host or at least prevent its reproduction rapidly.
6. It must reproduce / multiply rapidly.

Mode of action

Insects kill the weeds by boring into the plant to weaken / collapse the plant part and through consumption / destruction of vital plant parts.

Pathogenic organisms damage the host plant through.

a) Enzymatic degradation of cell constituents.
b) Production of toxins.
c) Disturbance of hormone system.
d) Obstruction in translocation of food materials and mineral nutrients.
e) Malfunctioning of physiological processes.

Effects of biological agents in weed management depends on:

1. **Choice of bio-agent:** The selected organism must be able to kill the weed or prevent its reproduction. It must locate host plant and should have high reproductive capacity sufficient to maintain the required population.
2. **Aggressiveness:** It is the competitive ability of bio-agents on weeds.
3. **Climatic conditions:** Climate affects the adoptability of the bio-agent and inturn the weed suppression.

Biological control agents

Insects

Weed	Bioagent	
1. *Lantana camera* L.	Larvae of *Crocidosema lantana* L. (flowers & fruits) Larvae of *Agromyza lantanae* L. (berries) Larvae of *Theclea echion* L. and *T. barochi* Lacewingbug: *Teleonemia scrupulosa* L..	
2. *Opuntia* sp.	*Cactoblastis cactorum* L. *Dacylopius tomemtosus* L.	*Dactylopius opuntiae* L. *Dactylopius indicus* L.
3. *Hypericum perforatium* (Europe, USA, Newzealand)	*Chrysalis hyperici* L. *Chrysalis gameliata* L.	*Agrilus hyperici* L.
4. *Cuscuta* sp.	*Melanogromyza cuscutae* L. *Smicronyx cuscutae* L.	
5. *Cyperus rotunudus*	*Bactra vermosana* L.	
6. *Imperata cylindrica*	*Orseoliella javanica* or *seoliella javanica* L.	
7. *Eichornia crassipes*	*Neochetina bruchi* L.	*Neochetina eichhorni* L.
8. *Parthenium hysterophorus*	*Zygogramma biocolorata* L.	

Plant pathogens: Disease causing pathogens *viz.,* Fungi, Bacteria, Virus offer greater promise for use as bio-herbicides. Bio-herbicides are the biological agents applied similar to the chemical herbicides in controlling weeds. The active ingredient in bio-herbicide is a living organism or its metabolites and is applied in moderate doses. Fungi are the most commonly used agents and hence, bio-herbicides are some time called as myco-herbicides.

S. No.	Product	Content	Target weed
1	Devine	A liquid suspension of fungal spores of *Phytophthora palmivora* causes root rot.	Strangle vine (*Morrenia odorata*) in citrus
2	Collego	Wettable powder containing fungal spores of *Colletotrichum gloeosporoides* causes stem and leaf blight	Joint vetch (*Aeschyomone virginica* L.) in rice, soybean
3	Bipolaris	A suspension of fungal spores of *Bipolaris sorghicola*	Jhonson grass (*Sorghum halepense*)
4	Biolophos	A microbial toxin produced as fermentation product of *Steptomyces hygroscopicus*	Non-specific, general vegetation

Fish: Common / grass carp (*Cyprimus carpio*) feed on grasses in the aquatic eco-system.

Mites: A spider mite (*tetranychus sp*) is found to be useful in controlling prickly pear.

Plants: Cowpea sown in between sorghum / pigeon pea rows effectively reduces the growth.

Mammels / animals: Many animals which feed weed as a fodder.

Merits of biological agents

1. Least harm to the environment.
2. No residual effect.
3. Relatively cheaper and comparatively long lasting effect.
4. Will not affect non-targeted plants and safer in usage.

Demerits of biological agents

1. Multiplication is costlier.
2. Control is very slow.
3. Very few host specific bio-agents are available at present.
4. No absolute guarantee of safety.

Chemical weed control: The use of chemicals generally referred as herbicides and used the control of weeds is called chemical weed control which offers greater potential in modern agriculture.

The word herbicide is derived from latin word **'Herba'** meaning 'Plant' and **'caedere'** meaning 'kill'.

History of herbicides: Common salts, ash *etc.,* have been used for centuries to control weeds on road side, fence rows and path ways.

Selective control of weeds in agriculture was observed in 1896 with the invention of Bordeaux mixture in France which is sprayed for control of downy mildew in grapes. It has damaged certain broad leaved weeds because of the presence of $CuSO_4$.

Between 1896 and 1908, several inorganic salts and acids were tried in small grains. The compound tested are sulphuric acid, carbon bisulphide, sodium arsenate, kaolinite, calcium cyanamide *etc.*

Between 1930 and 1940, boron compounds, thiocyanates, di nitro phenols, ammonium sulphate and other mineral salts were tried for weed management.

A real break through in selective chemical weed control was observed in 1994-95 with the discovery of 2.4-D (2,4-Di Chloro Phenoxy Acetic Acid) and MCPA (4-Chloro, 2-Methyl Phenoxy Acetic Acid) in USA and England respectively. These were found highly selective to cereals and phytotoxic to the broad leaved weeds. Also, they are non-corrosive, non-flammable and non-explosive.

Consequently, TCA, Dalapon, Chloro-Propham were discovered in 1945.

Today more than 300 herbicides in the world which were produced by about 100 companies.

Objective and scope of herbicide usage

Commercialization, industrialization and subsequent urbanization forced labour scarcity to agriculture sector. Indiscriminate use of inputs in the name of green revolution favoured the multiplication of weeds rapidly. Further, increased cost of hired labour, limited fuel & investment for mechanization, non-adoptability under extreme soil & climatic conditions for cultural and mechanical means are the reasons for introduction of herbicides in large scale. Hence, herbicides offer wider scope in weed management. However, herbicides are not aimed to substitute the physical, cultural or biological methods. Practical weed control is a combined effect of good cultural practices, herbicides and other techniques.

Merits of chemical weed control

1. Herbicide can be recommended for adverse soil and climatic conditions, as manual weeding is highly impossible during monsoon season.
2. Herbicide can control weeds even before they emerge from the soil so that crops can germinate and grow in completely weed-free environment at early stages.
3. Weed which resemble like crop in vegetative phase may escape in manual weeding. However, these weeds are controlled by herbicides.
4. Herbicide is highly suitable for broadcasting in closely spaced crops where intercultivation is impractical.
5. Controls the weeds without any injury to the root system of the associated standing crop especially in plantation crops like Tea and Coffee.
6. Reduces the need for pre planting tillage.
7. Controls many perennial weed species.
8. It is profitable where labour is scarce and expensive.
9. Suited best under minimum & zero tillage condition.
10. Highly economical.

Demerits of chemical weed control

1. Pollutes the environment and affects the soil micro flora and fauna.
2. It requires technical knowledge for calibration.

3. Some herbicide is highly costlier.
4. Suitable herbicides are not available for mixed and inter-cropping system.

Classification of herbicides

1. Based on Placement

i) **Soil applied herbicides:** Those herbicides which are applied on soil. They are mostly pre-emergence or pre-planting herbicides Ex. Butachlor, Fluchloralin.

ii) **Foliage applied herbicides:** Those herbicides, which are applied on the foliage of weeds. They are mostly post-emergence herbicides. Ex. 2,4-D, Glyphosate, Paraquat.

2. Based on Selectivity

i) **Selective herbicide:** Herbicide is considered as selective when in a mixed growth of plant species, it kills some species (weeds) without injuring the others (Crop). These are also called as narrow spectrum herbicides. Ex. Atrazine.

ii) **Non-selective herbicide:** It destroys majority of treated vegetation irrespective of crop and weed. These are also called as broad spectrum herbicides Ex. Paraquat.

3 Based on Mode of action / translocation

i) **Contact herbicide:** A contact herbicide kills those plant parts with which it comes in direct contact Ex. Paraquat.

ii) **Translocative / systemic herbicide:** Herbicide which tends to move from treated part to untreated areas through xylem / phloem depending on the nature of its molecule. The entire plant is poisoned and killed. Ex. Glyphosate.

4. Based on Time of application

i) **Pre-plant incorporation (PPI):** Application of herbicides before the crop is planted or sown. Soil application as well as foliar application is done here. For Ex. fluchloralin can be applied to soil and incorporated before sowing rainfed groundnut while glyphosate can be applied on the foliage of perennial weeds like *Cyperus rotundus* before planting of any crop. Ex. Trifluration, Fluchlorahin, EPTC.

ii) **Pre-emergence:** Application of herbicides soon after sowing before a crop or weed has emerged. In case of annual crops application is done after the sowing of the crop but before the emergence of weeds and this is referred as pre-emergence to the crop while in the case perennial crops it can be said as pre-emergence to weeds. For example soil application by spraying of atrazine on 3rd DAT to sugarcane can be termed as pre-emergence to cane crop while soil application by spraying the same immediately after a rain to control a new flush of weeds in a inter-cultivated orchard can be specified as pre-emergence to weed. Ex. Diuron, Metolachlor, Acetachlor, Oxyflurofen, Metribuzin.

iii) **Post-emergence:** Herbicide application after the emergence of crop or weed is referred as post-emergence application. When the weeds grow before the crop plants have emerged through the soil and are killed with a herbicide then it is called as early post-emergence. For example spraying 2,4-D Na salt to control parasitic weed striga in sugarcane is called as post-emergence while spraying of paraquat to control emerged weeds after 10-15 days after planting potato can be called as early post-emergence. Ex. Glufosinate, Paraquat, Lactofen, Quizalfop.

5. Based on the persistence / residual toxicity

i) **Residual herbicide:** A residual herbicide maintains its phytotoxic effect in soil for more than 3-4 weeks. Ex. 2,4-D., Triazine.

ii) **Non-Residual herbicide:** These are inactivated in the soil with in 3-4 weeks. Ex. Paraquot, Diquat, Glyphosate.

Based on chemical structure

a) **Inorganic herbicides:** These herbicides do not contain carbon atoms in their molecules. They are further classified into different groups based on their chemical nature.

b) **Organic herbicides:** These herbicides contain carbon atoms in their molecules. They may be oils or non-oils. Diesel oil, xylene type of aromatic oils, polycyclic aromatic oils are oil type of organic herbicides. Organic acids and salts as well as oil type of herbicides are not in use now. Majority of the present day herbicides are organic compounds, which are non-oils.

Herbicide formulations: Herbicides in their natural state may be solid, liquid, volatile, non-volatile, soluble or insoluble. Hence these have to be made in forms suitable and safe for their field use. Such forms are called as Herbicide formulations.

An herbicide formulation is prepared by the manufacturer by blending the active ingredient with substances like solvents, inert carriers, surfactants, stickers, stabilizers *etc.,* herbicides are formulated for easy handling and controlled activity on the target plants.

Need for preparing herbicide formulation

i) To have a product with physical properties suitable for use in a variety of types of application equipment and conditions.

ii) To prepare a product which is effective and economically feasible to use

iii) To prepare a product which is suitable for storage under local conditions.

Types of formulation

i) **Emulsifiable concentrates (EC):** A concentrated herbicide formulation containing organic solvent and adjuvants to facilitate emulsification with water. emulsifier helps in uniform distribution of the chemical in water and no stirring is necessary while spraying. Ex. 2,4-D ester, Nitrofen, Diallete.

ii) **Wettable powders (WP):** When the herbicides materials are of low solubility in water, they can be ground into fine powder for suspension in water. This type of formulations is called wettable powders. They require continuous agitation to prevent their settling and to give uniform concentration of the herbicides in the entire spray solution. Ex. Atrazine 80 % WP, Simzine 50 % WP, Diuron 80 % WP, lsorproturon 70 % WP.

iii) **Granules (G):** The inert material (carrier) is given a granular shape and the herbicide (active ingredient) is mixed with sand, clay, vermiculite, finely ground plant parts (ground corn cobs) as carrier material. Ex., Alachlor granules.

iv) **Water soluble concentrates (WSC):** Herbicide formulations that are in the form of soluble liquids are called water-soluble concentrates. E.g. 2,4-D amine, Dicamba, Diquat, Paraquat.

v) **Soluble powders (SP):** These are water-soluble powders. They form a homogenous solution when dissolved in water, which can be applied by spraying. Salts of most herbicides are soluble in water. Ex. Sodium salts of 2,4-D, TCA, Endothal, Dalapon.

vi) **Liquid suspension (LS):** If the active ingredient is not soluble in water, it is solubilised in organic solvents. When the product of active ingredient and solvent is added to water for spraying, it forms a liquid suspensions. These chemicals are comparatively cheaper than emulsifiable concentrates but require constant agitation during spraying to avoid setting. Ex. Atrazine, Cyprazine, Nitraline.

Methods of application

Factors influencing the methods of application are:

a) Weed-crop situation

b) Type of herbicides
c) Mode of action and selectivity
d) Environmental factors
e) Cost and convenience of application

Soil application of herbicides

a) **Surface application:** Soil active herbicides are applied uniformly on the surface of the soil either by spraying or by broadcasting. The applied herbicides are either left undisturbed or incorporated in to the soil. Incorporation is done to prevent the volatilization and photo decomposition of the herbicides. Ex. Fluchoralin-Left undisturbed under irrigated condition and Incorporated under rainfed condition.

b) **Subsurface application:** It is the application of herbicides in a concentrated band, about 7-10 cm below the soil surface for controlling perennial weeds. For this special type of nozzle is introduced below the soil under the cover of a sweep hood. Ex. Carbamate herbicides to control *Cyperus rotundus*. Nitralin herbicides to control *Convolvulus arvensis* L.

c) **Band application:** Application to a restricted band along the crop rows leaving an untreated band in the inter-rows. Later inter-rows are cultivated to remove the weeds. Saving in cost is possible here. For example when a 30 cm wide band of a herbicide applied over a crop rows that were spaced 90 cm apart, then two-third of the cost is saved. This method is useful where labour is expensive and inter-cultivation is possible.

d) **Fumigation:** Application of volatile chemicals in to confined spaces or in to the soil to produce gas that will destroy weed seeds is called fumigation. Herbicides used for fumigation are called as fumigants. These are good for killing perennial weeds and as well for eliminating weed seeds. Ex. Methyl bromide, Metham.

f) **Herbigation:** Application of herbicides with irrigation water both by surface and sprinkler systems. In western countries application of EPTC with sprinkler irrigation water is very common in Lucerne.

Foliar Application

i) **Blanket spray:** Uniform application of herbicides to standing crops without considering the location of the crop. Highly selective herbicides are applied in this method. Ex. Spraying 2,4-Ethyl Ester to rice three weeks after transplanting.

ii) **Directed spray:** Application of herbicides on weeds in between rows of crops by directing the spray only on weeds avoiding the crop. This could be possible by use of protective shield or hood. For example, spraying glyphosate in between rows of tapioca using hood to control *Cyperus rotundus* L.

iii) **Protected spray:** Applying non-selective herbicides on weeds by covering the crops which are wide spaced with polyethylene or metallic covers etc. This is expensive and laborious. However, farmers are using this technique for spraying glyphosate to control weeds in jasmine, cassava, banana.

iv) **Spot treatment:** It is usually done on small areas having serious weed infestation to kill it and to prevent its spread leaving the weed free gaps untreated. This can avoid possible wastage and reduces the cost. Rope wick applicator and Herbicide glove are useful here.

Methods of treating brush and trees

Brush weeds and unwanted trees are treated with herbicides with different methods depending on the situation :

1. **Foliage treatment:** is the most common method of treating the brush where the foliage is drenched with suitable herbicide. The treatment is best done when the brush leaves are fully expanded and growing actively.

2. **Bark treatment:** here the chemical is applied to the stem near the ground. It may be
   a) **Broadcast bark treatment:** Application of herbicide to the entire stem area of the plant. This is adopted when small woody plants having thick stem. Oil mixed phenoxy herbicides are applied in this method.
   b) **Basal bark treatment:** here the herbicides are applied to the stem base of 30-40 cm above ground, here selective trees can be killed without injuring other trees.
3. **Trunk injection:** Trunk injection often help the herbicide to penetrate through the bark. The trunk injections are made by frilling or notching with an axe or tree injector.
   a) **Frill treatment:** Here the herbicide is brushed or squirted to the cutting edge around the tree that are too large for bark treatment.
   b) **Notching:** The tree trunk may be notched with an axe for every 15 cm of the trunk circumference.
4 **Stump treatment:** Stumps of many species may sprout immediately after cutting. Here the trees are cut to the ground surface and concentrated herbicides / pastes are applied to the stump of cut surface.

Selectivity of herbicides and precautions in their use: The differential response of plants (Crops and Weeds) to herbicides is called as selectivity of herbicides. In other words, herbicides harm or kill weeds where as crop plants are unaffected due to the selectivity. The fundamental principle of herbicide selectivity is that more toxicant reaches the site of action in active form inside the targeted plants. This may be due to the difference in absorption, translocation, deactivation, carbon metabolism and resistance of protoplasm. The selectivity of herbicide may be due to one or combination of these processes.

1. Morphological and physiological differences observed between crop and weeds.
2. Man induces selectivity either by time or method of herbicide application or by other management practices.
3. Absorbents & antidotes may be used to prevent herbicide absorption by non-target plants.
4. Herbicides and their formulations may differ in their ability to contact the non-target plants.

I. Difference in absorption

1. **Foliage active herbicides:** The absorption of foliar applied herbicides primarily depends on retention of herbicide fluid on vegetation. The retention in turn depends on leaf properties like orientation, waxiness, pubescence, corrugation, ridges, depressions etc. In cereals like rice and wheat, have erect orientation than horizontal orientation of broad leaved weeds. Further, the active growing point is hidden in leaf whorls till booting stage where as dicotyledonous weeds have their sensitive growing points exposed to herbicides.
2. **Soil active herbicides:** The difference in uptake of soil applied herbicides is often based on differential interruption or availability. When herbicide is sprayed on the soil surface, it spreads into thin layer in the top 2-3 cm. most of the weeds germinate from shallow layer. The soil applied herbicides are toxic when they are absorbed by roots. Because of the bigger sized crop seeds, they are generally placed to a depth of 4-5 cm or more and roots develop deeper than 5 cm where there is no herbicide. As weeds germinate from the top layer they come in contact with herbicide and get killed. This type of selectivity is often called as **depth protection**.
3. **Induced selectivity:** Selectivity can be created or induced by using absorbents and antidotes. Absorbents are the materials which have the capacity to absorb the active chemical molecule

which are placed near the crop seeds to prevent the effect of chemicals. Activated charcoal adsorbs herbicides like 2,4-D, 2,4,5-T, Propham, butachlor *etc.* Germinating crop seeds and seedlings that are surrounded by a layer of activated charcoal are safe against these herbicides. Absorbents can be applied by different methods like applying over the seed row, in the seed hole, dipping the roots in charcoal-water mixture or seed pelleting.

Antidotes or safeners are the substances capable of antagonizing specific herbicide phyto-toxicity to plant. Some of the antidotes are NA (1,8-Nepthalmic anhydride) & R-25788 are proven effective antidotes of EPTC *etc.*

II. Differential translocation

There are instances where equal amounts of herbicides absorbed by plants, but translocated at different rates. The selectivity between sugarcane (Tolerant) and beans (Susceptible) to 2,4-D is due to low translocation in sugarcane and rapid translocation in beans. This selectivity further depends on water status and activity of the plants.

III. Differential protoplasmic resistance

Application of herbicides causes deficiency of certain vitamins, amino acids or other constituents. Protoplasm of certain plants can resist the deficiency and tolerate to herbicides. For instance, plants that show tolerance to dalapon can withstand Pentothenic acid deficiency in their tissue and also resist precipitation of their cell proteins.

IV. Differential rate of deactivation

The differential deactivation of herbicide in the plants induces selectivity. The deactivation may be due to metabolism, reverse metabolism or conjugation:

1. **Metabolism:** Breakdown of herbicide inside the plant into non-toxic metabolites is known as metabolism of herbicides. Selectivity of terbacil between tolerant peppermint and susceptible morning glory is due to differential rates of breakdown. Peppermint metabolises terbacil rapidly at the cost of photosynthesis while herbicide persist long time in morning glory due to poor metabolism of terbacil.

2. **Reverse metabolism:** In some of the metabolic reactions of herbicides, the intermediate chemical substances are more toxic than their parent compound. The herbicides 2,4-DB & MCPB are not phytotoxic originally and metabolism of these compounds through enzymatic β-oxidation form a toxic compound which is not found in leguminous plants and are resistant to these compounds.

3. **Conjugation:** Coupling of intact herbicide molecules with some of the plant cell constituents in living plant is known as conjugation. Conjugation takes the toxic herbicide concentration out of main stream of activity in plants. Chloramben is rapidly conjugated in roots of tolerant soybean with glucose molecule to form N-glucosyl chloramben. In susceptible plants, conjugation is slow and herbicide enter to the site of action. Tolerance of grasses to 2,4-D is also due to conjugation.

V. Differential carbon metabolism

Herbicides affect carbon metabolism and metabolic activity differently in different plants. Chlorofenprop methyl decrease the reducing sugars gradually in wild oats while barley is not affected.

Precautions in storage and handling of herbicides

1. Store the herbicides away from fertilizers, seeds, food, feed and children in cool and ventilated place. The left over herbicide must be retained in originally labeled container. The container must be airtight to avoid caking, oxidation, fuming *etc.*

2. Transportation of herbicides should be done with trolley and not with bare hand / body.
3. Prepare the herbicide dilutions in open spaces away from the source of irrigation water. Wear rubber gloves, a pair of eye glasses and cover the nose with a cloth.
4. Mix the solution with a stick and not by bare hand. Bury the empty containers deep in any waste land.
5. Keep fresh water and soap handy to meet any emergency.
6. Don't smoke or eat during and in between the spray intervals. Also don't use mouth to blow the clogged nozzles.
7. Take bath and wash the cloths thoroughly as soon as the spray is over.
8. In case of accidental oral intake, induce quick vomiting by any local methods.
9. Avoid herbicide drifts from reaching the non-target plant and animals.

Integrated weed management (IWM)

Any single method for weed management mayn't be effective in controlling weeds. Integration of different methods like sanitation, mechanical, cultural, biological and chemical means kept the weeds under check at an economic cost. IWM uses a variety of technologies in a single weed management with the objective to produce optimum crop yield at a minimum cost taking in to consideration ecological and socio-economic constraints under a given agro-ecosystem.

IWM is a system in which fibeor more methods are combined to control a weed.

Need for integrated weed management

a) No single weed management practice is effective in controlling wide range of weed flora.

b) Continuous use of same herbicide creates resistance or causes shift in the flora.

c) Indiscriminate herbicide use and its effects on the environment and human health.

Advantages of integrated weed manaement

1. It shifts the crop-weed competition in favour of crop.
2. Prevents weed shift towards perennial nature.
3. Prevents resistance in weeds to herbicides.
4. No environmental pollution and no danger of herbicide residue in soil or plant.

Herbicide resistance: Herbicide resistance is the inherited ability of a plant to survive and reproduce following exposure to a dose of herbicide that would normally be lethal to the wild type. In a plant, resistance may occur naturally due to selection or it may be induced through such techniques as genetic engineering. Resistance may occur on plants as result of random and infrequent mutations. Through selection, susceptible plants killed and resistant plants survive to reproduce without competition from susceptible plants. If the herbicide is continuously used, resistant plans successfully reproduce and become dominant in the population.

Herbicides are active at one or more target sites within a plant. Target sites are enzymes, proteins or other places in the plants where herbicides bind and thereby disrupt normal plant functions.

Parameters to study weed control effect

1. Weed Control Efficiences (WCE)

It is the reduction in weed population due to weed control method over unweeded check.

It gives magnitude of reduction in weed dry weight.

$$WCE = \frac{(WD_C - WD_T)}{WD_C} \times 100$$

WD_C = weed denisity in control plot (number m^2)

WD_T = weed density in treated plot (m^2)

2. Weed Index (WI)

It is the measure of the crop yield loss occurred across treatments in comparison to a weed free plot adopted in the experiment.

$$W_I = \frac{(Y_{WF} - Y_T)}{Y_{WF}} \times 100$$

Y_{WF} = Crop yield in weed free plot

Y_T = Crop yield in treatment plot

3. Weed Smothering Efficieness (WSE)

It is usually worked out in inter cropping systems to know the effect of inter crop on weed control or weed smothening.

$$WSE = \frac{(M_{dw} - I_{dw})}{M_{dw}} \times 100$$

M_{dw} = Average dry weight of weeds in main crop

I_{dw} = Average dry weight of weeds in intercropping situation.

Allelopathy: The term allellopathy was coined by Hans Molisch (1937), which derived from Greek word allen of each other and pathos to suffer. Allelopathy is the detrimental effects of chemicals or exudates produced by one (living) plant species on the germination, growth or development of another plant species (or even microorganisms) sharing the same habitat.

Allelo chemicals are produced by plants as end products, by-products and metabolites liberated from the plants; they belong to phenolic acids, flavanoides and other aromatic compounds *viz.*, terpenoids, steroids, alkaloids and organic cyanides.

Allelopathic effect of weeds on crops

1. Maize

1. Leaves and inflorescence of *Parthenium* sp. affect the germination and seedling growth.
2. Tubers of *Cyperus esculentus* affect the dry matter production.

2. Sorghum

1. Stem of *Solanum* sp. affects germination and seedling growth.
2. Leaves and inflorescence of *Parthenium* affect germination and seedling growth.

3. Wheat

1. Seeds of wild oat affect germination and early seedling growth.
2. Leaves of *Parthenium* affects general growth.
3. Tubers of *C. rotundus* affect dry matter production.
4. Green and dried leaves of *Argemone mexicana* affect germination & seedling growth.

4. Sunflower

1. Seeds of *Datura* affect germination & growth.

Allelopathic effect of crop plants on weeds

1. Root exudation of maize inhibits the growth of *Chenopodium album.*
2. The cold water extracts of wheat straw when applied to weeds reduce germination and growth of *Abutilon* sp.

Allelopathic effect of weeds on weeds

1. Extract of leaf leachate of decaying leaves of *Polygonum* contains flavonoides which are toxic to germination, root and hypocotyls growth of weeds like *Amaranthus spinosus.*

2. Inhibitor secreted by decaying rhizomes of *Sorghum halepense* affect the growth of *Digitaria sanguinalis* and *Amaranthus* sp.
3. Exudates of *Cassia* sp. affect germination & growth of *Parthenium hysterophorus*.

References

Mandal, R.C., 1993, Weed, weedicide and weed control, Agro Botanical Publishers, New Delhi.

Palaniappan, S.P. 1985, Cropping Systems in Tropics-Principles and Practices, Willey Eastern Ltd. New Delhi.

Rana, S.S. and Rana, M.C. 2011, Cropping Systems, Department of Agronomy, College of Agriculture, CSK, Himachal Pradesh Krishi Vishvavidyalaya, Palampur.

Reddy, S.R., 1999, Principles of Agronomy, Kalyani Publishers, Ludhiana, India.

Reddy, S.R. and Reddy, APK, 2016, Fundamentals of Agronomy and Agromeorology, Kalyani Publishers, Ludhiana, India.

Yellamanda Reddy and Sankara Reddy, 2016, principles of agronomy, Kalyani Publishers, Ludhiana, India.

Annexure

1. National Institutions for Agricultural Research

1. Central Arid Zone Research Institute (CAZRI), Jodhpur, Rajasthan.
2. Central Institute for Cotton Research (CICR), Nagpur, Maharastra.
3. Central Institute of Agricultural Engineering (CIAE), Bhopal, Madhya Pradesh.
4. Central Institute of Fisheries Technology (CIFT), Cochin, Kerala.
5. Central Marine Fisheries Research Institute (CMFRI), Ernakulam, Cochin, Kerala.
6. Central Plantation Crops Research Institute (CPCRI), Kasaragod, Kerala.
7. Central Potato Research Institute (CPRI) Simla. Himachal Pradesh.
8. Central Research Institute for Dryland Agriculture, (CRIDA), Hyderabad, Andhra Pradesh.
9. Central Research Institute for Jute and Allied Fibres (CRIJAF), Barrackpore, West Bengal.
10. Central Rice Research Institute (CRRI) Cuttack, Orissa.
11. Central Soil and Water Conservation Research and Training Institute. (CSWCRTI), Dehradun, Uttarakhand.
14. Central Soil Salinity Research Institute (CSSRI), Karnal, Haryana.

15. Central Tobacco Research Institute (CTRI) Rajamundry. Andhra Pradesh.
16. Central Tuber Crops Research Institute (CTCRI), Thiruvananthapuram, Kerala.
17. Indian Agricultural Research Institute (IARI), New Delhi.
18. Indian Agricultural Statistics Research Institute (IASRI), Pusa, New Delhi.
19. India Grassland and Fodder Research Institute (IGFRI) Jhansi, Uttar Pradesh.
20. Indian Institute of Horticultural Research (IIHR), Hassaraghatta, Bangalore, Karnataka.
21. Indian Institute of Pulses Research (IIPR), Kanpur, Uttar Pradesh.
22. Indian Institute of Soil Science (IISS) Bhopal, Madhya Pradesh.
23. Indian Institute of Spices Research (IISR), Calicut, Kerala.
24. Indian Institute of Sugarcane Research (IISR), Lucknow, Uttar Pradesh.
25. Indian Lac Research Institute (ILRI), Ranchi, Jharkhand.
26. Indian Veterinary Research Institute, (IVRI) Izat nagar, Uttar Pradesh.
27. Jute Technological Research Laboratories (JTRL), Calcutta, West Bengal.
28. National Bureau of Plant Genetic Resources (NBPGR), IARI Pusa Campus, New Delhi.
29. National Bureau of Soil Survey and Land Use Planning, (NBSS & LUP) Nagpur, Maharashtra.
30. National Dairy Research Institute (NDRI), Karnal, Haryana.
31. National Research Centre for Agroforestry (NRCAF), Jhansi, Uttar Pradesh.
32. National Research Centre for Banana (NRCB), Tiruchirappalli, Tamil Nadu.

33. National Research Centre for Oil Palm (NRCOP), West godavari District, AP.
34. National Research Centre for Weed Science (NRCWS) Jabalpur, Madhya Pradesh.
35. Sugarcane Breeding Institute, (SBI) Coimbatore, Tamil Nadu.

II. International Institutions for Agricultural Research

1. Centro International de Agricultura Tropical (CIAT) Columbia.
2. Central International de la Papa (CIP) (International Institute of Potato) Lima, Peru.
3. Centro International de Mejoramiento de Maiz Y Trigo (CIMMYT) (International Centre for maize and Wheat Improvement) Londres Mexico, D.F.Mexico.
4. International Centre for Agricultural Research in the Dry Areas (ICARDA), Syria.
5. International Crops Research Institute for the Semi Arid Tropics (ICRISAT), Hydcrabad.
6. International Food Policy Research Institute (IFPRI), Washington, D.C. U.S.A.
7. International Institute of Tropical Agriculture (IITA) Nigeria.
8. International Laboratory for Research in Animal Diseases (ILRAD), Nairobi, Kenya.
9. International Service in National Agricultural Research (ISNAR), Netherlands.
10. International Livestock Centre in Africa (ILCA), Addis Ababa, Ethiopia.
11. West Africa Rice Development Association (WARDA) Ivory coast, Liberia.
12. International Rice Research Institute (IRRI), Manila, Philippines.

13. Asian Vegetable Research and Development Centre (AVRDC) P.O.Box 42, Shanhau, Tasnan 941, Taiwan, Republic of China.
15. International Centre of Insect Physiology and Ecology (ICIPE) Nairobi, Kenya.
16. International Council for Research in Agro Forestry. (ICRAF) Nairobi, Kenya.
17. International Irrigation Management Institute (IIMI), Colombo, SriLanka.
18. Consultative Group on International Agricultural Research (CGIAR), N.W.Washington. D.C.U.S.A.

III. Year of Establishment of Institutes

1947 - CTRI at Rajamundry (Tobacco).

1949 - CPRI at Patna.

1956 - CPRI shifted to Simla.

1950 - IARI established at New Delhi.

1952 - IISR at Lucknow (Sugarcane).

1955 - NDRI at Karnal (Dairy).

1956 - PIRRCOM Project for intensification of regional research on cotton, oilseeds and millets. (Central Cotton Research Institute – Regional Centre).

1959 - CAZRI at Jodhpur (Rajasthan) (Arid zone).

1960 - IRRI, Los Bonas, Phillippines.

1962 - IGFRI at Jhansi Uttar Pradesh; G.B. Pant Nagar Agricultural University and technology at Pant Nagar.

1963 - CTCRI, Trivandrum (Tuber crops).

1965 - IAAP (Intensive Agriculture Area Programme).

1969 - CSSRI (Central soil salinity Research Institute) at Karnal (Haryana).

1970 - CPCRI at Kasargod (Kerala) (Plantation crops) / Drought Prone.

1972 - ICRISAT at Patancheru, Hyderabad.

Zeitfracht Medien GmbH
Ferdinand-Jühlke-Straße 7
99095 Erfurt, Deutschland
produktsicherheit@kolibri360.de